Third Edition

ROSE LORE

ROSE LORE:
ESSAYS IN CULTURAL HISTORY
AND SEMIOTICS

Frankie Pauling Hutton,
Editor

UnCUT/VOICES Press

ISBN: 9 783981 386349

Bibliographic information published by the Deutsche Nationalbibliothek.
The Deutsche Nationalbibliothek lists this publication in the Deutsche
Nationalbibliografie; detailed bibliographic data are available on the Internet at
http://dnb.d-nb.de
Frankfurt am Main: UnCUT/VOICES Press, 2015
Original: Frankie Hutton, ed. *Rose Lore. Essays in Cultural History and Semiotics.* Lanham, MD: Lexington Books. A Division of Rowman & Littlefield Publishers, Inc., 2008.

UnCUT/VOICES PRESS
Martin Luther Str. 35, 60389 Frankfurt am Main, Germany
Tobe.levin@uncutvoices.com www.uncutvoices.com
Geschäftsnummer HRB 86527, U.G. Haftungsbeschränkt

A note on variety in documentation styles: considering the many nationalities involved in this project, we have not insisted on conformity in notes, footnotes, or bibliographic entries but prefer to reflect the diversity of backgrounds from which we come. For scholarly pursuit, however, all vital information is included.

Front cover design by Uma M. Swaminathan and Sreedevi Kashi who wrote:
"The mystical red rose provides a concept that encompasses all matter and space – when the cosmos resonates to Shiva's drums, and the dance to annihilate worldly sorrows releases souls from the bondage of illusion in preparation for the new creation."
Back cover art by Martin Cervantes, painting commissioned especially for this edition.
Layout: Steffen Schenk, Frankfurt am Main

Also from UnCUT/VOICES PRESS

Khady with Marie-Thérèse Cuny. *Blood Stains. A Child of Africa Reclaims Her Human Rights.* Trans. Tobe Levin. ISBN: 978-3-9813863-0-1
A ground-breaking memoir by Senegalese Khady, Europe's leading activist against female genital mutilation, forced and early marriage, and unequal gender relations in the African Diaspora. "Khady's account ... is wrenching and necessary reading." Henry Louis Gates, Jr., Harvard University

Hubert Prolongeau. *Undoing FGM. Pierre Foldes, the Surgeon Who Restores the Clitoris.* Foreword by Bernard Kouchner, founder of Doctors without Borders and former Foreign Minister of France. Trans. and Afterword by Tobe Levin. ISBN: 978-3-9813863-1-8.
"Can excision be reversed? Can the wounded sex be healed? French surgeon Pierre Foldes ... return[s] to eager patients their sensitivity, femininity, and courage to break the chain. A must read!" – Elfriede Jelinek, 2004 Nobel Laureate in Literature

Nick Hadikwa Mwaluko. *WAAFRIKA. Kenya. 1992. Two Women Fall in Love.* Foreword by Ginni Stern. 978-3-9813863-2-5
"Powerful and fearless, ... Mwaluko's ground-breaking drama challenges our local and global imaginings of African love and sexual identity." – Tracie Jones, Harvard Graduate School of Education

Tobe Levin, ed. *Waging Empathy. Alice Walker, Possessing the Secret of Joy and the Global Movement to Ban FGM.* ISBN: 978-3-9813863-3-2
"With enthusiasm I endorse these essays from around the world showing that readers and critics alike appreciate a story about FGM. Our stories will end the scourge once and for all." – Soraya Miré, filmmaker and author.

Rose Lore:
Essays in Cultural History and Semiotics
Edited by Frankie Hutton

Table of Contents

Preface to the 2015 Third English Edition

Amazingly, an eventful seven years have passed since the original hardcover *Rose Lore* was published with Lexington Books, a division of Rowman & Littlefield, Inc. Since then, UnCUT/VOICES Press provided a paperback which showcased the work of Dr. Gamze Demiral on the rose in Turkish culture. Meanwhile, I've given many lectures in New Jersey, Florida, Maryland and Wiesbaden near Frankfurt, Germany; and because there is no other endeavor quite like it, we've managed to launch the rose project whose mission is to educate the global public about the cultural, historical and metaphysical qualities of the earth's quintessential flower. It was a feat that could not have been done alone; the right people have crossed my path to aid me in ways that may never be understood fully by all who read this. At this writing, that is, aspects of the rose remain partly veiled and phenomenal; our team's discoveries are fully intended for everyone, but each must "get up to get it," so to speak. Put another way, for most of us, getting to know the rose requires work.

After my rose lecture in Germany, I met artist Martin Cervantez who approached to tell me the talk had struck a resonant note with him. Of course, authors love to ride the incredible spirit behind such comments because they are so galvanizing. Martin volunteered to paint a rose in honor of the rose project and went on to do a contemporary, magnificent, larger-than-life rendition that garnishes our back cover; it is partnered with a more traditional Hindu view of the flower by Uma Swaminathan and her daughter Sreedevi on the front.

What's more, I connected with Ryan Dunne and his talented, then bride-to-be Catherine Yard. We bonded; they performed as a highlight of one of my rose lectures, and Ryan, a musician and actor with strong attachments to Irish culture, has written a chapter on the rose in Irish music for this edition. I could say more. However, even if a few people connected with the book and the rose project remain off the screen, their well-wishing, research, helpful comments and aid remain poignantly appreciated. In Florida, Natasha Reeder and Lillian Mayer have been long-haul friends of the Rose Project.[1]

In addition to our new essay on the 'Irish' rose, this volume is enhanced by three Asia-related chapters by Vicki Eaton, born in Vietnam; Uma Swaminathan of South Indian heritage; and Yaping Qian of China who teaches translation at China Women's University, Beijing. In fact, inspired by Cathy's research, the manuscript was taken on by Ding (Dylan) Zhanggang and the Center for Translation Studies at China Women's University, submitted for publication and approved by prestigious Peking University Press with scheduled appearance in Mandarin in 2015. And so the wide connections made possible by the rose continue in this earthly realm and, in fact, into infinity.

Like earlier editions, this one is dedicated to the memory of my parents, Frank and Georgia Nettles Pauling, and also to Evan and Emily who accompanied me to lectures in 2013 in Deerfield Beach and Miami, Florida, and provided capable assistance which was divinely inspired, I am sure. They were both, – get this –, eight years old at the time.

Frankie Pauling Hutton, Ph.D.
www.roseproject.com

References

[1] See www.roseproject.com. Retrieved 14 January 2015.

ROSE LORE

By Maria Jaschok

'… in a literary world where the rose is seen archetypally, all things have human form' - so says Hazard Adams. As the rose allows itself to be anthropomorphized, it allows us thus to breathe human meaning into a nonhuman part of the physical world that may be considered our common heritage. Firmly anchored in a shared and idealized imaginary without the artificial borders and demarcations that separate humanity, our common heritage, we might argue, also expands. The rich and multifaceted manifestations of human creativity – a creativity which has enabled women and men of different ages and places to mythologize, name, interpret, classify, sensualize and also, and here lies moral ambiguity, consume the rose – are beautifully illustrated in the varied contributions to this anthology. The ROSE LORE assembles a seductive tapestry of cross-cultural richness and discovery. When individual contributors speak to the object of the rose in diverse discursive and socio-cultural contexts, they are addressing important parts of themselves, allowing us, the readers, thereby the discovery of our own natures.

Yet whilst the rose as the universe's 'quintessential archetypal flower' may be seen as a fundamental symbol of the Jungian Collective Unconscious, immediately recognizable as a universal representation of beauty and perfection, the writings provide testimony to the poignancy of Difference and Otherness. The rose lore of beauty and perfection, moreover, is pervaded by a lore which speaks of dissonances and ambiguities engendered by injustice, inequality and outright destruction. As intimation of the loss of human innocence, 'the sick rose' becomes thus a potent, multifaceted symbol, mirroring the complexities of our interdependent and ever-conflicted world.

Sensitively translated from the original English into Chinese, the participants in this wonderful project of translation are to be congratulated for making this revised anthology accessible to Chinese readers. It gains greatly from a new preface and three chapters from South India, Viet Nam, and China.

Dr. Maria Jaschok
International Gender Studies at Lady Margaret Hall
University of Oxford

Preface to the 2012 Special Edition

To reach a wider audience, we are privileged in this 2012 special soft-cover, limited edition of *Rose Lore: Essays in Cultural History and Semiotics* to have UnCUT/VOICES Press as our publisher.

A student of the Gurdjieff tradition, I have quietly taken note of what Gurdjieff had to say about roses in Beelzebub's "Tales to His Grandson" from the First Series of *All and Everything* (page 210). "In short, both the climate of this country and everything else so delighted the hunters and their families that none of them, as I have already said, had any desire to return to the continent Atlantis, and from that time on they remained there, and soon adapting themselves to everything, multiplied and existed, as is said 'on a bed of roses'." Some of us know the real meaning of this quote. Translation: we on earth are so steeped in avarice and warmongering that many have forgotten how to love and be easy with each other; at times we seem to be sinking slowly like the ill-fated continent of Atlantis. ...

With this new edition, we are reminded again that the rose is no ordinary bloom but, instead, the most important flower in the universe with much to teach us about how to live and love peacefully and richly in this life and beyond.

We ought never to stop working for the shift in consciousness it represents. This is the message, the nectar of the major world religions, although fundamentalists and others in self-service have missed and continue to miss the point of living. In this regard, it is also our boon to have a new chapter written by Gamze Demirel, a Turkish scholar affiliated with Suleyman Sah University in Istanbul who gives us a glimpse of the rose in her rich cultural tradition and classical Ottoman Empire history which includes not only Qur'anic insights but shows us how, over decades, many Turkish people have embraced the rose in their everyday lives. Indeed, we thank Fatma Mizikaci, a dear, gentle friend and fine scholar, for her assistance in bringing the research of Dr. Demirel to our attention.

Since introducing the first edition of *Rose Lore*, the Rose Project has been launched and has begun to speak for itself carrying on the work inspired by this book, softly, gracefully, up and down the eastern seaboard, from Palm Beach to Baltimore to New York. With gratitude to the Rose Work Team quietly planting the seeds to accomplish our rose work, we are especially appreciative of the diverse support of all who have been inspiring and helpful, including but not limited to Toni and Carol Randolph, actor David Keltz, violinist Airi Yoshioko, Meg Ventrudo, Cathy Lowe, Barbara Bell Coleman, Vernon O'Meally, Daryl Dance and significant others, some from far corners of the earth including Fisher Island, Hiroshima, Istanbul and New York. We have all been nurtured in understanding the simple, elegant, infinite rose.

Frankie Pauling Hutton, Ph.D.
www.roseproject.com

Prologue

A cosmic mandala in the shape of a shining white rose was revealed to Dante in a vision. I learned this from Aniela Jaffe, C.G. Jung's secretary and pupil speaking in *Man and His Symbols*. In the many years since reading Jung, I have come across the rose as an icon many times. Why does this special flower serve so well? Perhaps it has a perfect shape that invokes a mandala of wholeness; perhaps we are attracted to its vibrant color and variety of hues; perhaps it is simple beauty that stirs something in our hearts, but whatever the reason, the rose as a symbol has become embedded in our human psyche.

In this collection, remarkable essays explore rose symbology. I propose that you, the reader, attend to whatever special meanings resonate for you, and allow that sensitivity to lead in pursuit of your star. Then, as the mystic Gurdjieff might have said, you will have embarked on a search for meaning that shall be nothing less than just roses, roses.

Seymour B. Ginsburg, Esq.
Boca Raton, Florida

Foreword

By Mario D. Fenyo

Symbols can be extremely powerful. There are animate and inanimate objects, animals, plants, celestial bodies and "things" that have assumed global symbolic significance, sometimes in excess of their intrinsic importance. The exception may be the Sun, the life-giver that is an almost equally powerful divinity for many cultures around the globe, sometimes equivalent in power to its actual status in our solar system. The Moon is another divinity, another all-powerful symbol, appreciated everywhere. Other celestial bodies, three or four of the planets in our system, and a few "fixed" stars such as Sirius are powerful symbols in their own right.

There are many animal symbols, ranging from fish in general to fishes in particular, to birds of all feathers, to insects and snakes. Some members of the animal kingdom have a symbolic meaning extending far beyond the region they inhabit, for example the lion, naturalized in Great Britain and many other places.

There are inanimate symbols that have metamorphosed into animate by virtue of their symbolic power or vice-versa. Such is the case of certain numbers, of certain Arabic numerals, or of metals, the most obvious being gold and silver. Religious symbols are all-pervasive: here the symbol denotes a reality of an order far beyond (or above) the importance of the signifier.[1] Obviously, among the religious symbols there is the rose. Among the more abstract symbols we find historic figures and events, which often acquire a parallel history as symbols and then, in turn, influence us, quite out of proportion with their impact as analyzed by the "scientific" "objective" historian. They make up the mythology of modern nations.[2]

In the plant kingdom, perhaps the most important symbol of all is the Tree which, in addition to its almost infinite manifestations, has come to symbolize life. Then there are the flowers which are, botanically rather than symbolically, female sexual organs. If the Sun is symbolically male, or a phallus, it follows that flowers are female, symbolically as well botanically. "Thus with the rose, earlier the flower of Venus, then the flower of the Virgin Mary, from the cup of which many sun heroes, as well as the divine 'rose' child are born."[3]

Of course, all flowers function biologically as female organs, but the rose is the most popular floral symbol of the vulva.

The rose is a grateful plant. She will grow almost anywhere, in almost any soil, in almost any climate: she has been observed across the African continent, in the Far East, in the tropical gardens of the Caribbean and South America, and all over Europe. She will grow in the garden, and she will grow wild. While her thorns may prick you, – maybe because she is democratic and a lover of freedom – she likes to please all passers-by, much as those who once sacrificed themselves on the steps of the temple of Ishtar. She wants to be enjoyed rather than appropriated. She is not demanding. She will open up to the sunlight, to anyone who looks at her, even to the Little Prince,[4] who searched the globe to find a companion for the lonely rose on his personal asteroid. Voyaging from planet to planet, meeting cranks and people obsessed with money and numbers, it took him a while to realize that down here, on Earth, roses are beautiful but common. Too bad Saint-Exupéry was a racist!

Frankie Hutton has other fine, award-winning works to her credit, but from now on she will be recognized, first and foremost, as the author who has given the Rose more of its due, for it has many dues as an archetypal symbol. This book is a collection, thus the credit has to be shared by those who have contributed to the bouquet – they have recognized and conveyed, along with the poets and the artists, the transcendental, universal values of the symbolism in this super-flower.

References

[1] Eliade, Mircea, ed. *The Encyclopedia of Religion*. Vol.14. NY: Macmillan, 1987. 206.

[2] See, for instance, Andras Gero. *Imagined History*. NY: Columbia UP, 2006.

[3] Marcell Jankovics. *Book of the Sun*. Trans. Mario Fenyo. NY: Columbia UP, 2001.

[4] Antoine de Saint-Exupéry. *The Little Prince*, Chapter XX. Trans. Katherine Woods. NY: Reynal & Hitchcock, 1943.

Introduction

Chapters in this collection are faithful to the indomitable spirit and staying power embodied in the world's most perfect flower, the rose. If semiotics is the study of signs and symbols, then the rose rightfully takes her place at the center of this discipline, for it has much to teach us by its example of quiet perfection, beauty, and all-around usefulness to humankind.

You may, however, wonder which precise rose of the myriads blooming in innumerable gardens concerns us.

In one of its many books simply called *The Rose*,[1] the American Rose Society contends that it is counterproductive to look for specific varieties of roses in literary or artistic works because such attempts "invariably lead to speculation." Rose symbolism can, moreover, be readily appreciated "without knowing what specific rose, if any, was intended." The chapters in this collection reflect this: it doesn't matter precisely what type of rose comes to mind in order to admire the flower as one of the universe's archetypical symbols. Nonetheless, it is the genre commonly known as the red tea rose that drives this anthology.

And as we shall see, when studied intently, the rose can be viewed as a providential if recondite and seemingly impenetrable messenger from the universe to humans. Though not imperceptible, its true missives are well veiled and not for the profane. No doubt that is as it should be.

The essays here are not therefore about the fragile, cliché-filled Valentine as it has been overly-commercialized. Still, the most compelling flower on earth has indeed found its place symbolically in numerous paintings, family crests, heraldry, coins, flags, dinnerware, clothing, perfume, fabric design, architecture and even furniture. Hundreds of thousands of females are named for the flower worldwide. Songs and movies, paintings, poetry and books too numerous to mention have the rose in their titles. One example by Eva Rosenkranz, *The Romance of the Rose*[2] celebrates in painting and verse the flower's great variety of uses and should not go unnoted. From a Jewish mystical, metaphysical perspective, another book, Adin Steinssaltz's *The Thirteen Petalled Rose* offers classic discourse on the essence of mitzvoth or obligations as enhanced by the rose in this religious faith.

Similarly, the essays in *Rose Lore* celebrate the rose's varied qualities and applications, its message inherent in the suppleness with which it serves others. Arcane, cultural and historical aspects of the flower's special mission in the universe are woven into essays presented here for the first time in aggregate.

Some inkling of the rose's special errand can be gleaned from the ease with which the flower is crossbred into lovely varieties. The rose itself works hard but quietly for the universe and for humanity, and there is a powerful message in the fact that it does. Providing gifts such as wonderful scent, nutrients and glamour, its numerous botanical hybrids are remarkable to touch, to smell and even to taste. In this way it conveys the news that each human being must learn flexibility, beauty and the inclination to serve and connect with others.

Hence, the rose, primordial, botanically perfect and vast in its adaptability as a symbol, is user-friendly from a number of perspectives, some of which are not well known but will be unfolded in this collection.

Looking at roses from around the globe, Orietta Sala's comprehensive illustrated guide to 180 varieties notes that the flower's presence had been recorded 2000 years before Christ, and that the bloom first "appeared on Earth during the Tertiary era, as indicated by the many fossil roses found in ... the world – [in] the Baltic region, India, Oregon and Colorado." Sala notes 150 species of roses, 95 indigenous to Asia and 18 to North America. Two essays in this volume confirm the flower's longevity; indeed, it had appeared on earth long before the so-called Tertiary era. In fact, the American Rose Association provides evidence of fossil records that have tentatively identified roses as 30 to 40 million years old. No doubt, the flower predates the arrival of human beings as we know them today. For instance, judging by illustrations in the cosmological, anthropological pictographs of *Oasphe*, a little known bible of the late nineteenth century that presents insights stretching back for many thousands of years, the rose appeared in hieroglyphics of Atlantis. Imbedded in mysticism, culture and metaphysics, the rose has, for eons, offered knowledge and understanding of itself as far more than a fragrant, lovely flower with a thorny stem. It has been embraced by secret orders, fraternities, and social clubs.

Although we need not master the horticultural specifics of myriad types, it is useful to consider them. Thus, while *The American Rose Society Encyclopedia of Roses* edited by Charles and Brigid Quest-Ritson lists 2000 of the "best rose species and cultivars" available worldwide, the "rose expert" Dr. D. G. Hessayon divides the floral delight into seven groups: shrub roses, climbers and ramblers, patio roses, floribunda roses, miniature roses, ground cover roses and, of course, the popular hybrid tea rose.[3] A world-wide favorite, the red rose comes from this latter lot. And perhaps not entirely fortuitous, Dr. Hessayon groups the botanical flora into seven major types, a number that has great significance in metaphysics.

Although now, as a result of cross-breeding, we have an almost endless variety of roses, as a symbol the red rose is clearly the most revealing of the secrets of the universe. Above us, nearly out of sight, is the most provocative evidence of its importance: the rose nebula[4] known by NASA (the United States' National Aeronautical and Space Administration) as NGC 2244. Four thousand five hundred light years away in the cosmos, this vibrant nebula hovers over the earth in the shape of a red rose as if underscoring messages of the flower from above. It brings to mind: "as above, so below." Invisible to the naked eye, the rose nebula can be seen only with the aid of a powerful telescope. Conversely, on earth the botanical rose is at once lovely to look upon and to touch but has a thorny stem as well, like life itself.

Called simply the "red rose," it benefits humankind in many practical, common ways, for instance, in rose hip tea, splendid jellies and jams. Diets may call for rose hips often partnered with vitamin C. Bulgarian rose oil is even an ingredient in natural progesterone gel.[5]

What could be more divine than a flower that is fragrant, edible, hardy, and beautiful with curative qualities as well as cultural and metaphysical staying power reaching back beyond recorded history? It seems there's still another message in its centuries of survival. Although a certain revelatory knowledge came to him too late to save his own family, the great 16th century physician and clairvoyant Nostradamus revealed that the rose had vast preventive, medicinal powers that could be used to protect humans from the Black Plague.

Nostradamus wasn't the first to be cognizant of the rose's special therapeutic qualities, however, for in Native American tribal culture as well as in Hinduism and traditions on the Indian sub-continent, rose nectar is widely known and the rose itself used not merely as a symbol of beauty and Spirit for thousands of years but also for its healing properties.

What's more, worldwide, a number of sacred, obscure documents reiterate the rose's talismanic meanings. In its entire splendor, the rose is becoming better known today, but it has always been available as a divine symbol through which those who are awake and aware can understand their true place and mission in the universe. As an enduring, providential icon, the red rose bespeaks the work each individual can and must do to connect with Cosmic Consciousness. Hindu mystics knew that symbolism is divine language, providing the "keyholes to doors in the walls of space. It is through these keyholes that we can reach the Qabalistic light and with it draw back bolts that hold domus santi spiritus into eternity." In these essays, we see that this light is possible through living the messages of the rose.

The chapters in *Rose Lore*, heartfelt and flowing from diverse perspectives, cover the rose's little known history, culture, symbolism and mysticism. Although its commercialization makes the flower easy to take for granted, a few individuals in all epochs and geographies have known that, beyond botanics, the rose is a magical and especially winsome plant. The Phoenix, edited by Manly Hall and thought by many to be the "bible" of early theosophy, reminded readers in its 1931-32 edition that "When a nation ceases to serve the beautiful, it has already begun to die: when a cause departs from the truth, that cause has already failed."[6] In all its manifestations, the rose is synonymous with beauty, and it is likely not coincidental that the anagram of rose is Eros, the name for the Greek God of love.

The meaning for the universe in the beauty of the rose has been understood by rare accolites in multiple philosophies: the flower appears as an important, quiet symbol in Qabalistic teachings; in fraternal secret orders such as the Order of the Golden Dawn; in Rosicrucianism; in H. Blavatsky's *Secret Doctrine;* under veil in the Christian *Holy Bible;* in *The Emerald Tablets;* in *Oasphe* and more. Although remaining obscure in these tomes, it is nonetheless profound, as indicated by an Atlantean pictograph. Furthermore, even if few Catholics know the origins and history of the rosary, the Catholic Church has historically used roses in its rosaries, undoubtedly an important icon.

A significant theme in some chapters is the number of great individuals who have embraced the rose. It's unusual and heretofore unrecorded in aggregate that a bevy of world class men and women were "given" the rose as a symbol in their work or encountered it as a central, sometimes urgent sign in their own lives. To illustrate, at various times in Europe, Nostradamus, William Butler Yeats, Rudolph Borchardt and H. Blavatsky knew the rose. In the United States John Ballou Newbrough, C.A. Miles and Edward Bach used it. In Japan, Dr. Tomin Harada was inspired by roses after treating badly disfigured victims of the atomic bomb in 1945. From the gulag of Soviet Russia, Daniel Andreev adopted the rose as an important, urgent symbol in his mystical, social philosophy. The singular work of all of these achievers has represented, in most cases, unimaginable feats that no ordinary human beings could accomplish. How did the rose touch the lives of so many great, divergent individuals from far corners of the globe, from Russia to France, from New Jersey, USA, to Hiroshima, Japan? These people hailed from different walks of life, and the work they undertook differed greatly, but all shared the inspiration of the rose in their dedication and service to humanity.

To all who have worked on this collection, it has become increasingly clear that the rose is far more than a lovely fragrant flower with thorny stems. Tbe thorns, too, have symbolic import, conveying that some of the rose's messages must be carefully considered and guarded, as several chapters underscore. Silence, service and sincerity are crucial for the rose to appear fully in one's life.

Each chapter in this collection uncovers rare truths. Each also stands alone in its approach to the rose as a symbol on earth and, when mystical implications are considered, even beyond. Just as the rose nebula is the ever-present reminder in the cosmos of what the botanical rose on earth reveals, there is much work to be done but too few involved in doing it. Why? The flower requires higher awareness if its full splendor and deep messages are to appear. Only discipline, sincere service without expectation of reward, and behavior in keeping with the spirit of the rose will bring the light of connection to others.

Each contributor provides a unique vision of the flower. My colleague, professor of history in the University System of Maryland at Bowie State University and translator of Marcell Jankovics's *Book of the Sun*, Mario Fenyo wrote the reflective Foreword. Chapter one was penned by Michael Price, a member of the Wikwemikong First Nations who teaches biology and plant science at a tribal college in Northern Minnesota; he reveals that many American Indians – among over 500 tribal affiliations and a population of about two million – knew the power and spirit of the rose long before Whites. A research scholar affiliated with Harvard and the University of Oxford, Tobe Levin, a humanitarian and collegiate professor emerita of UMUC - Europe, shows how 'cutting the rose' has come to represent the movement to stop female genital mutilation. Lisa Cucciniello, author of the third chapter, grew up in the Catholic tradition; a teacher, independent researcher and graduate student, she traces the origins of the rose in the Catholic rosary. Our fourth is by Albert Amao. Of Peruvian heritage, he now lives in the United States as a psychotherapist and teaches metaphysics in the tradition of the ancient Qaballah. The fifth chapter is by Monika Joshi, a registered nurse and former officer in the Indian Army. Born in Bombay/Mumbai, Joshi is a licensed Ayurvedic practitioner now residing in California. She reveals the East Indian reverence and appreciation for the rose and notes its presence in healing and in Hindu culture. Montgomery Taylor, author of the sixth chapter, is a former university professor, master astrologer, author and musician who brings a unique rendition to the collection: astrology and the rose. Offering invaluable editorial assistance and translation expertise have been Alexandr Prodovikov, a former Russian economist and Moscow resident; and world-class translator Hisae Ogawa of Osaka, Japan. Both assisted with my chapter "Black Plague to Gulag" by providing translation expertise on great individuals from their countries.

And in 2012, with this limited paperback edition, it is my delight to include the research of Gamze Demirel, a Turkish scholar who enlightens us on aspects of the rose in Turkish - Islamic tradition and literature.

In sum, there is so much to learn about the rose and all its esoteric properties. Beyond the scope of this collection are the mysteries surrounding the ways some of us have met, interacted and shared rose stories while coming to know the unique rose as the universe's quintessential, archetypical flower. Gracefully, we offer this book to all who study mysticism, semiotics, herbal medicine and cultural history, for the rose is "planted" in all of these disciplines and beyond.

Frankie Pauling Hutton
New Jersey 2007; Revised 2012
www.roseproject.com

References

[1] The American Rose Society, Beth Smiley, Ray Rogers, eds. *The Rose.* (NY: Dorling Kindersley, 2000). 3.

[2] Eva Rosenkranz. *The Romance of the Rose: A Celebration in Painting and Verse.* (NY: Prestel Publishers, 2005).

[3] D. G. Hessayon. *The Rose Expert.* (NY: Sterling Publishing, 2002). 7.

[4] Also known as NGC 2244, the rose nebula can be seen with the aid of a powerful telescope. It is said to be a stardust sculpture containing globules of dark dust and gas that are slowly being eroded by nearby massive stars. The rose nebula has been noted in Jerry T. Bonnell and Robert J. Nemiroff, eds. *Astronomy 365 Days: The Best of the Astronomy Pictures of the Day Website.* (NY: Abrams Publisher, 2006). Also of interest regarding the rose nebula is *The New Atlas of the Stars* by Axel Mellinger.

[5] See www.NUTRACEUTICAL.COM Retrieved 12 January 2015.

[6] Manly Hall. "Introduction." *The Phoenix, An Illustrated Review of Occultism and Philosophy.* (Los Angeles, CA: Hall Publishing, 1931-1932 Edition). 5.

1

Wild Roses and Native Americans

By Michael Wassegijig Price

I begin this chapter with a story from the Native American oral tradition that tells of the significance of wild roses to the Anishinaabe, the Indigenous peoples of the Great Lakes:

Long ago, the Anishinaabe people were starving after an extended summer drought. The berries were few and scarce, the caribou had moved farther north, and the fish had vanished from the lakes and rivers. Soon, sickness and starvation were upon the village.

One day, a group of young warriors came across Makwa – Black Bear – walking through the forest. The eldest of the young warriors decided that they should kill the bear and bring him back to the village for a feast. In doing so, the young men would be revered as great hunters during this time of need.

As the young warriors crept up on Makwa, the eldest noticed that he seemed to be looking for something. Quietly, they began watching his behavior. Makwa began pulling over long slender thorny bushes and eating the red fruit that was fixed to them. The young warriors watched patiently, waiting for a good shot, as Makwa gorged on the red berries.

Next, Makwa proceeded toward the lake. Feeling a little tense, the youngest of the warriors felt that they should kill Makwa before he got away. As they began to affix their arrows onto the bear, they noticed that he was pulling up the tall water grass and eating the white fleshy bulb beneath the surface of the water. Makwa ate the bulbs while the young warriors lurked in the bushes.

After he had gotten his fill, Makwa lumbered off into the forest and disappeared. The young warriors quickly ran over to the bushes and ate both the red berries and the whitish bulbs of the tall water grass. They ran back to the village and told the elders what they had witnessed.

From that moment on, Makwa, the bear, was no longer regarded as a food source, but as the teacher of herbs and medicines. Medicine men and women of the Anishinaabe people revere the black bear for its knowledge and wisdom.[1]

The red berries atop the thorny bushes are rose hips, and the white fleshy bulbs of the tall water grass are cattails. This story reflects the in-depth knowledge that Native Americans possess about the plants and animals around them and the pathways by which they obtained such knowledge. The Anishinaabe, the Native Americans of the Great Lakes region of the United States, have long known that rose hips are the final feast of black bears before they hibernate for the winter. The bears gorge themselves on the fruits in the late autumn when the temperatures begin to fall. Soon afterwards, the bears and their cubs retreat to their dens where they sleep until late April.[2] To the Anishinaabe, herbal knowledge obtained from the bears was considered sacred knowledge.[3] Through observing the behavior of bears in the forest, Native Americans learned that rose hips can be a food source.

This is the story of the wild rose and its connection to the indigenous peoples of North America. Native Americans have many legends, medicinal remedies and ceremonies that are centuries old relating to the many species of wild rose. This thorny shrub is known as a scant food source, herbal medicine and good luck charm. Wild rose can be found across the prairies of the Dakotas, the highlands of Washington, the Smoky Mountains of North Carolina and Tennessee, and the northern woodlands of Minnesota and Ontario. Wild rose has many colorful and descriptive names within the different indigenous languages. Historically, Native American tribes have found many uses for this perennial herb, which we will discuss in this chapter.

Wild Roses, Trade Beads and Tribal Insignia

Many Native American tribes can be identified by their traditional designs and ornamentation. While many plains tribes, such as the Lakota, Dakota and Nakota, used geometric designs in ornamentation, many northern woodland tribes, such as the Anishinaabe and Iroquois in the northeastern United States, and the Cree in central Canada, use floral designs which identify them as woodlands people. Other tribal nations that utilize floral designs are the Blackfeet and Okanagan of the northern high plains, and the Athabascans of interior Alaska. In the early 18th century, European traders brought glass beads and velvet cloth with them and traded with the tribes in exchange for furs. But even before the introduction of glass beads, many tribes had already developed intricate floral patterns and utilized items such as dyed porcupine quills, bone beads, and colorful cordage to decorate their outfits. The introduction of glass beads quickly transformed the appearance of traditional regalia for many tribal communities who traded with the European explorers. By the late 1700s, Anishinaabe people had already incorporated vibrant colorful beadwork and black velveteen cloth into their traditional attire.[4]

The Anishinaabe use woodland floral designs to ornament their dance outfits, bandolier bags, cradleboards, and various other ceremonial objects. The floral designs reflect the local plant species with which they were most familiar. The most popular designs for Ojibwe beadwork are wild roses, rosebuds, bluebells, maple leaves, acorns, and wild grapes, and are used as much today as they were over two hundred years ago.[5]

The author of this article has sat in on conversations with Anishinaabe elders who stated that originally the floral designs represented medicinal plants. Wearing the design of a particular medicinal plant could bring power and good medicine to that individual.[6] These floral designs can be seen today at powwows and ceremonial gatherings all across Anishinaabe Country.

Wild Roses in Native American Languages

Native American languages are as diverse as the landscape of Turtle Island – a metaphor for the North American continent meaning that the land is actually on the back of a giant turtle. Many indigenous tongues share a root with many dialects. The major language groups across the North American continent are Siouan, Algonquin, Tewa, Diné and Iroquoian.

In general, Native American languages tend to rely heavily upon metaphoric and symbolic relationships. Many descriptions and abstract concepts are woven together using elements of the local environment and cultural traditions relative to each tribal community. One study attempted to categorize the indigenous naming of wild plants, but found that the indigenous ways of knowing plants were far different from the scientific Linnaeus method of categorization.[7]

The Anishinaabe People comprise many Native American tribes including Ojibwe, Ottawa, Pottawatomie, Menominee, Meskwaki, and Shawnee. The language of the Anishinaabe is placed by linguists in the Algonquin class, a group that includes many tribal peoples from Minnesota to Maine, from Indiana to Hudson's Bay, Ontario.

In the Ojibwe language, spoken by the Anishinaabe people, the name for wild rose is *Oginiiminagaawanzh* (pronounced O-GINEE-MINA-GAW-WUNZH). An analysis of the word reveals:

ogini – his mother; *imin* – fruit; *agaawanzh* – small bush

The translation is *"Mother Fruit from a Small Bush."* The name refers to edible fruits, or rose hips, that are found at the tips of the rose stalks. This name also refers to several species including Virginia Rose *(Rosa virginiana)*, Prickly Wild Rose *(Rosa acicularis)*, Smooth Rose *(Rosa blanda)* and Swamp Rose *(Rosa palustris)*. All of these species are indigenous to the northern boreal forests and peat lands of the Great Lakes region.[8, 9]

The Anishinaabe word, Ogin, has a more contemporary translation which means *"tomato."* Tomato and rose hips are both edible fruits from each respective shrub. The multiple meanings of this word make translation complex and challenge understanding the naming of this flowering plant.[10, 11]

The Meskwaki people are a small tribe of Anishinaabe that reside in central Iowa. Meskwaki translates to *"People of the Red Earth,"* possibly referring to the deep red, iron-rich clay soils of the region. These southern relatives of the Anishinaabe call the wild rose *Kishipi'iminaki* which means "to itch like hemorrhoids" referring to the bowel irritation caused by ingesting the microscopic hairs found on the seeds within the rose hip fruit.[12]

The Anishinaabe have a different name for the Wild Prairie Rose *(Rosa arkansana)* which is named *Bizhikiwiginiig* (pronounced BIZHIKI-WIG-INEEG). An analysis of the word reveals:

bizhiki – buffalo; *gin* or *ogin* – rose flower; *iig* – plural ending

The translation means "Buffalo Roses." Buffalo once roamed the tall grass prairies of the high plains. Their hooves tilled the soil and created landscapes suitable for prairie grassland species. The buffalo and grasslands evolved together over the centuries. But rarely did the buffalo venture into the eastern woodlands of Minnesota and Manitoba. The Anishinaabe associated the Wild Prairie Rose (*Rosa arkansana*) with the buffalo's natural habitat.[13] Lakota tribal member and Professor of Bison Rangeland Science at Little Hoop Tribal Community College, James Garrett, Ph.D., says that the buffalo graze on the wild rose leaves, stalks and flowers throughout the summer and autumn, and rose hips throughout the winter months.[14]

The tribes of the Great Plains speak a different language from the Anishinaabe People of the woodlands. The Plains tribes have a strong connection to the tall grass prairies and the buffalo. Tribes of the Great Plains such as the Dakota, Omaha, Pawnee, Cheyenne and Blackfeet all have names for the various rose species in their respective languages:[15, 16]

Tribe	Tribal Name	Latin Classification
Dakota	Onzhinzhintka	*Rosa pratincola*
Ponca-Omaha	Wazhide	*Rosa pratincola*
Pawnee	Pahatu	*Rosa pratincola*
Blackfeet	Kinii	*Rosa acicularis*

The presence of wild roses on the plains inspired the naming of the Rosebud Sioux Reservation in South Dakota. Through the Indian Reorganization Act of 1934, Rosebud Agency was established in Kyle, South Dakota, to administer governmental services to the Sicangu Band of Lakota. The name "Sicangu" (pronounced SEE-CHONG-GOO) is a Lakota word which means "burnt thighs." The Rosebud Sioux reservation got its name from the charismatic Lakota leader Spotted Tail and the not-so charismatic governmental officials who visited the region in 1877 and chose it as the site of the new Lakota territory. Spotted Tail and other Lakota noticed the prevalence of wild roses growing across the prairie landscape. The wild roses were reminiscent of the Rosebud River Country in present-day Montana, where they traditionally camped while en route to the Little Big Horn Country. The name "Rosebud" memorialized a part of their history to which they could never return.[17]

Wild Roses and Native American Spirituality

Native American spirituality is a complex weaving of the physical world with the spiritual world. It recognizes that all living things possess a spirit, including animals, plants, rocks and rivers. By recognizing that these things have a spirit, Native peoples developed ways to commune with the spirit world that surrounds them. These relationships have evolved throughout centuries of harvesting foods, conducting ceremonies, practicing traditions, and storytelling.

The spiritual landscape of Native Americans is as diverse as the physical landscape that they inhabit. Ceremonies, herbal knowledge and cultural practices are directly related to the Eco region and homeland of a particular tribe. The ceremonies practiced by Pueblo peoples of the desert southwest would have little meaning to the woodland people of the northern woodlands, and vice versa. Buffalo ceremonies conducted by the plains tribes were practically unknown to the Cherokees of the Smoky Mountains. Though many tribal nations share certain customs and beliefs, describing the spirituality of Native Americans as one group is difficult, if not impossible.

Many traditional tribal peoples believe that wild rose has a unique spirit that is deserving of respect and recognition. Before any part of the plant is harvested, permission is asked from the spirit of that particular plant, and a gift is usually given in gratitude for its harvest. Tobacco is usually a gift given to appease the spirits. By performing this ceremony, Native peoples humble themselves and take only what is needed for their use. This worldview has sustained indigenous communities with their environments for centuries.

The meaning and purpose of many ceremonies were never shared with Western society. When the U.S. Government in conjunction with Christian churches attempted to force assimilate young Native children into English-speaking Christian society in the early 20th century, many ceremonies and tribal traditions went underground and were held in secrecy for generations. Some ethnographers and anthropologists recorded tribal languages and customs, but one key point to remember is that their writings were interpreted through the eyes of the ethnographer, not by the indigenous peoples who created them. For instance, the Omaha practiced a ceremony by rubbing crushed rose petals on their hair. There were many assumptions and theories as to the importance of this practice, and many assumed that it was used as a hair perfume. The Omaha never shared the purpose of this ceremony with the outside world and the rite continues to be held in secrecy.[18]

For many Native American tribes, smoking was a ceremonial activity that created kinship and goodwill between people, communities and nations. Pipe ceremonies were conducted between peoples seeking relationship or common alliances. Ceremonial pipes were carved from the sacred red pipestone *(Catlinite)* found only in the area of Pipestone, Minnesota. Tribal peoples made smoking mixtures from different plants indigenous to their homelands. The Plains tribes such as the Omaha, Pawnee, and Dakota smoked the inner stem bark of wild rose in their pipes, or they mixed it with the inner bark of red willow and tobacco.[19] The Klamath people of Oregon used wild rose stems as pipe stems because they became rigid when dried and could be easily hollowed out.[20]

The Okanagan-Colville people proudly called themselves "The People of the Plateau" because their ancestral homelands are located in the eastern highlands of Washington State. Okanagan spiritual leaders made good luck charms by boiling wild rose stems in water and then sprinkling the brew over fishing lines and nets for a successful catch. Hunters sprinkled the water charm on themselves for good luck which also served to diminish human scent for hunting. This water charm was also used to chase away bad spirits or prevent bad medicine inflicted by another person. In addition, Okanagan medicine men would use the branches of wild rose to sweep out a freshly dug grave to prepare it for burial ceremonies.[21]

Wild Rose as Food Source
Every part of the wild rose plant, which includes stems, roots, leaves, hips and buds, was used as either a food source or food additive. Rose hips were harvested in the autumn or throughout winter. Hips were gathered after the first frost because freezing temperatures made the fleshy tissues tender, though the ratio of fleshy fruit to seeds is small. Though not a staple food, rose hips were considered "famine food" because they could be found above the snow during the winter months when starvation was most prevalent. The leaves and flower petals are eaten raw throughout the growing season, while the roots, stems, leaves, petals and hips were all used in teas.[22]

Many tribes knew the food value of rose hips, but they were also aware that the fine microscopic hairs on the seeds could create bowel irritation. A coyote story shared by many tribal communities warns of eating too many rose hips:

> *Once Coyote ate the bright red fruits of the wild rose. The next day, his anus began to itch. It itched so much that Coyote was desperate to seek relief. Coyote grabbed a handful of willow bushes and began wiping himself. He wiped himself so much that he began to bleed, and he bled all over the willow branches. Still, he could not stop the itching. That is why the branches of the Red Willow (Salix laevigata) are bright red.*[23]

Anishinaabe elder and scholar Basil Johnston tells a story of how Nanaboozhoo, the Trickster god of the woodlands peoples, gave thorns to the wild rose to protect itself from the gluttonous rabbits. For nearly decimating the wild roses from the landscape, the animals ganged up on rabbit, grabbed him by the ears which stretched them out long, and punched him in the mouth which split his upper lip.[24] This story demonstrates the importance of the wild rose as a food source for other animal species, and the physical characteristics of the rabbit are reminiscent of this legend.

Other relative species of the wild rose (Family: Rosaceae) which are traditional food sources for Native Americans are Blackberries, Raspberries and Smooth Juneberry. The Ojibwe names are as follows:[25]

Common Name	Ojibwe Name	Latin Classification
Blackberry	Odatagaagomin	*Rubus allegheniensis*
Raspberry	Miskomin	*Rubus idaeus*
Smooth Juneberry	Gozigwaakomin	*Amelanchier laevis*

Wild Rose as a Medicine

Wild Rose contains significant amounts of vitamins C and A, minerals such as calcium, phosphorus and iron, and it can act as an antioxidant. "Wild Rose is the super-plant of the Blackfeet people," states Roslyn LaPier, Blackfeet tribal member, Tribal Ethnobotanist and faculty member at the Piegan Institute in Browning, Montana. "Every part of the plant is used: roots, leaves, flowers, hips and buds."[26] Rose species can also act as an antiseptic, astringent and diuretic. Throughout the centuries, tribal communities shared their medicinal remedies, while other groups used the plant in entirely different ways.

Many Native Americans, including woodlands and plains tribes, used poultices of various wild rose species for eye mucous irritation and cataract inflammation. People of the northern tribes such as the Ojibwe, Cree, and Iroquois occasionally suffered from snow blindness and eye soreness during the winter months. Healers from these communities created an eye wash from the petals, stem bark and root bark of wild rose to relieve eye pain and irritation.[27] The Ojibwe, in particular, treated cataracts by creating two eye washes, one made from the root bark of wild rose and the other from the root bark of wild raspberries, and applied them separately and successively. The wild rose eye wash treated the irritation, and the wild raspberry eyewash worked to heal the tissues around the eye. This eye-wash treatment was administered three times a day until the eyes healed.[28]

Teas were also used to treat diarrhea and gastrointestinal problems. The Meskwaki people of northern Iowa created a decoction of the fruit to treat hemorrhoid inflammation, although the microscopic hairs from the rose hip seeds had to be carefully removed because, as mentioned earlier, it too would cause bowel irritation.[29] The Blackfeet of northwestern Montana made necklaces with dried rose hips and used a concoction of wild rose stems and root bark to treat children and adults with diarrhea.[30] The Anishinaabe people used dried rose petals as a medicine to treat heartburn.[31]

In the Plateau region of Washington State and British Columbia, Prickly Rose *(Rosa acicularis)* and Dwarf Rose *(Rosa gymnocarpa)* grow abundantly. Okanagan people call wild rose "Coyote Berry" because they observed the coyotes eating rose hips. For bee stings, Okanagan healers placed chewed leaves on the sting area to reduce swelling and irritation of the skin.[32] People of the Shuswap First Nations (Thompson Indians) in British Columbia treated athlete's foot by placing rose leaves in their moccasins.[33]

The ancestral homelands of the Cherokee people are in the Smoky Mountains of western North Carolina and Tennessee. Here, the Virginia Rose *(Rosa virginiana)* is the predominant wild rose species. The name for wild rose in the Cherokee language is *Jisdu Unigisdi* which translates to "what the rabbits eat." Cherokee healers created a "medicine bath" using the wild rose roots to treat babies and children with worms.[34]

Many Native American tribes used wild rose in the treatment of open or bleeding flesh wounds. The Ojibwe created a decoction of pulverized wild rose roots to prevent bleeding by applying it directly to the wound. The Crow of central Montana created a vapor using crushed roots in boiling water to treat nosebleeds. They also used the same remedy to make a compress to reduce swelling of skin wounds.[35] For treating burns, the Pawnee harvested the large hypertrophied outgrowths on the rose stem called galls. Galls are tumors that form on the stem as the result of a bacterial infestation. Healers would char the wild rose galls in a fire, crush them into a poultice, and apply it directly to the wound.[36] In a similar fashion, the Paiutes of the Warm Springs Reservation in Oregon chewed the wild rose galls into a poultice and applied it to boils.[37]

The spirit and essence of wild roses have permeated Native American cultures for centuries. This small thorny shrub has inspired healers to cure the ill, quelled hunger during times of struggle, created funny stories and fables that would live for generations, nourished the body, and lifted the spirits of the oppressed. Like the wild rose, Native Americans have persevered in the harshest of environments, adapted to rapidly changing surroundings, survived adverse conditions, and yet have maintained their beauty and essence. The wild rose possesses much power, magnificence and elegance. Human beings have only to open their minds and souls to discover its medicine.

This article is dedicated to my mother, Rita Wassegijig (Bright Sky), who passed on to the spirit world on May 5, 2002. Since childhood, her favorite flower had been the wild rose.

References

[1] Michael Wassegijig Price, recitation of an Ojibwe story from the oral tradition. There are several versions of the same story that are shared by many Anishinaabe communities.

[2] Harriet V. Kuhnlein, *Traditional Plant Foods of Canadian Indigenous Peoples: Nutrition, Botany and Use* (Netherlands: Gordon and Breach, 1991) 248-9.

[3] Frances Densmore, *How Indians Use Wild Plants for Food, Medicine, & Crafts* (Washington: United States Government Printing Office, 1928; rpt. NY: Dover, 1974) 324.

[4] Carrie Lyford, *Ojibwa Crafts*. (Wisconsin: R. Schneider, 1943, rpt. 1982) 145.

[5] Ibid. 147.

[6] Michael Wassegijig Price, recollection of casual conversations with women tribal elders from the Leech Lake Band of Ojibwe in northern Minnesota. The passing of knowledge from one person to another, a centuries-old custom, is known as the "oral tradition."

[7] Iain J. Davidson-Hunt et al. "Iskatewizaagegan (Shoal Lake) Plant Knowledge: An Anishinaabe (Ojibway) Ethnobotany of Northwestern Ontario." *Journal of Ethnobiology* 25(2) (Fall/Winter 2005) 189.

[8] James E. Meeker et al. *Plants Used by the Great Lakes Ojibwa* (Wisconsin: Great Lakes Indian Fish & Wildlife Commission, 1993) 82, 225, 394.

[9] John Eastman, *The Book of Swamp and Bog: Trees, Shrubs, and Wildflowers of Eastern Freshwater Wetlands* (Pennsylvania: Stackpole Books, 1995) 155-8.

[10] Fredric Baraga, *A Dictionary of the Ojibway Language* (St. Paul: Borealis/ Minnesota Historical Society Press, 1878) 317.

[11] John D. Nichols and Earl Nyholm, *A Concise Dictionary of Minnesota Ojibwe* (Minneapolis: U. of Minnesota, 1995) 238.

[12] Huron Smith, *Ethnobotany of the Meskwaki Indians* (Milwaukee: Bulletin of the Public Museum of the City of Milwaukee, 1928) 242.

[13] Meeker, ibid. 53.

[14] James Garrett, Interview by the author. Rapid City, South Dakota. March 25, 2007.

[15] Melvin R. Gilmore, *Uses of Plants by the Indians of the Missouri River Region* (Washington: U.S. Government Printing Office, 1919; rpt. Lincoln: U. of Nebraska, 1977) 33-4.

[16] Roslyn LaPier, Interview by the author. Browning, Montana. March 15, 2007.

[17] James Rattlingleaf, Interview by the author. Kyle, South Dakota. April 23, 2007.

[18] Melvin Gilmore, *A Study in the Ethnobotany of the Omaha Indians* (Nebraska State Historical Society Collections, 1913) 314-357.

[19] Gilmore, *Uses of Plants.* 33-4.

[20] Frederick V. Coville, *Notes on the Plants Used by the Klamath Indians of Oregon* (Contributions from the U. S. National Herbarium, 1897) 87-110.

[21] Turner et al. *Ethnobotany.* 131.

[22] Kuhnlein, Ibid. 248-9.

[23] Michael Wassegijig Price, recitation of a traditional story shared by many Native American communities. There are several versions that exist for this particular theme.

[24] Basil Johnston, *Ojibway Heritage* (Toronto: McClelland and Stewart, 1976; rpt. 1990). 44-5.

[25] Meeker, Ibid. 29, 125, 231.

[26] Rosyln LaPier, Interview by author. Browning, Montana. March 2007.

[27] Kelly Kindscher, *Medicinal Wild Plants of the Prairie* (Lawrence, Kansas: University Press of Kansas, 1992) 189-93.

[28] Densmore, Ibid. 292.

[29] John Eastman, *The Book of Swamp and Bog: Trees, Shrubs, and Wildflowers of Eastern Freshwater Wetlands* (Pennsylvania: Stackpole Books, 1995) 155-8.

[30] Kindscher, Ibid. 189-93.

[31] Smith, Ibid. 385.

[32] Turner, Ibid. 131.

[33] Ibid. 267.

[34] Linda Averill Turner, *Plants Used as Curatives by Certain Southeastern Tribes* (Cambridge: Botanical Museum of Harvard University, 1940). 29.

[35] Jeff Hart, *Montana Native Plants and Early Peoples* (Helena: Montana Historical Society Press, 1976; rpt. Montana Historical Society and Montana Bicentennial Administration, 1992). 35-6.

[36] Gilmore, Ibid. 33-4.

[37] James Michael Mahar, "Ethnobotany of the Oregon Paiutes of the Warm Springs Indian Reservation" (M.A. Thesis, Reed College, 1953) 25.

2

FGM –

Or Cutting the Rose in Alice Walker's Garden

By Tobe Levin

> Many have written of genital mutilation, and many have denounced
> it [including many respectful of tradition]. ... I feel enfeebled here,
> unable to add a constructive ... insight, depressed by the persistence
> of a repulsive "rite," and made small, as we all are, by capitulation
> through inertia. Genital cutting is an extreme abuse of human rights.
> Like slavery and apartheid, it is unacceptable. How can we stop it?
> By talking about it with angry, unbitten tongues. By never forgetting
> about it, and by not letting the issue slide back into obscurity now
> that we have learned of its pervasiveness and tenacity.
>
> **Natalie Angier.**
> *Woman. An Intimate Geography.*[1]

In "The Cut," Maryam Sheikh Abdi courageously tells us what it
was like the day she faced the knife.[2] As the children were taken one
by one, the nauseous six-year-old waited with "ears blocked" by a
single sound, the shrieks and sobs of "wailing ... girls."

Then, it was her turn.

> Obediently, I sat between the legs of the woman who would
> hold my upper abdomen, and each of the other four ...
> grasped my legs and hands. I stretched apart [with] each limb
> firmly held [until], under the shade of a tree ... the cutter
> began ... [ellipses in original]

"To this day," Maryam proclaims decades later, the wound remains
fresh: she had screamed until her "voice grew hoarse, [and no] cries
could come ... as the excruciating pain ate ... flesh."

As in *Born in the Big Rains. A Memoir of Somalia and Survival*,
where Fadumo Korn, in a near death experience, floats above a
scene of carnage, Maryam faints only to awaken to "unbearable"

"agon[y]" as well as "a slap across [the] face" for cowardice. She had never ceased writhing; and even though the surgery itself – the cutting and stitching – was over, "the pain kept coming in waves, each wave more pronounced than the one before it ... for the blood oozed and flowed," attracting the "scavenger birds [who circled above] and perch[ed] on nearby trees."

What happens next? Remarkable for its programmatic character, the narrative poem spells out in unwonted detail what the 'initiate' endures. After "the bloody sand" is "scrubbed" off, a hole is dug, and the *malmal* is pounded. It will be "pasted where [the] severed vaginal lips had been" before the legs are tied. Over a fire containing "dried donkey waste" and herbs, Maryam sits, only to hear "the blood dr[i]pping on the charcoal." Will she bleed to death?

Bound in camel hide, she survives and is instructed how to sit, stand, and shuffle so as not to disturb the "sealing" of "that place," now a source of endless torment. Excretion is hell. As she lies on her side, the urine, "more burning than ... the razor," leaks out drop by drop. Unwashed and undried for hours without end, the wound flames.

The healing takes a month during which time the diet is strict: no vegetables, oil, or meat, and very little water. Nor do the victims bathe. Hence, the lice attack. Nesting "between the ropes and our skin," the insects bite; the wounds itch. Respite? There is none.

Nura Abdi, together with her age-mates, goes through this, too. After losing consciousness, she recovers to see

> blood on the floor and ... what had been sawed off all of us... tossed ... in a pile. .... Later I learned that someone had dug a hole and buried [those parts] somewhere in the courtyard. Exactly where, we were never to learn. "What do you need to know for?" was all they would say. "It's long gone to where it belongs. Under the earth."[3]

With exceptional courage, Maryam, Fadumo and Nura testify to their infibulations, a genital assault that culminates in burial of the "offensive" body parts. Out of sight, out of mind.

Cutting the Rose
Although the late Monique Wittig, in a post-modern quip, claimed to have had no vagina, thereby removing herself from the class of

candidates for genital clipping, individuals whom patriarchal binaries identify as clitoris-carriers too often undergo ritual sexual attack. The figures are well-known and increasingly horrible because failing to decrease significantly.[4] Despite campaigns that began in the 1970s, official estimates of victims still cite more than 130,000,000 who have undergone or are threatened with varying types of amputation (15% at risk of infibulation, the remaining 85% of clitoridectomy or excision).[5]

Several questions obviously follow. Why are female genitals so treated, without reciprocity?[6] Why is the custom so tenacious? And is this 'merely' an African problem?

More than 200 years ago, English Romantic poet William Blake captured centuries of hostility to women's sexuality in "The Sick Rose":

> O Rose, thou art sick!
> The Invisible worm,
> That flies in the night
> In the howling storm,
>
> Has found out thy bed
> Of Crimson joy;
> And his dark secret love
> Does thy life destroy.[7]

A metaphor for female genitalia, the fragile flower signals pain and decay throughout Western literary history. Superficial and transient in its splendor, the rose attracts the worm, a vile and sepulchral penile metonymy, implying lethal male desire linked to fear of sex. While consummation in the "howling storm" endangers both participants, the male's *angst* translates into violence, and he neuters the temptress. Faced with a treacherous organ, seductive yet malodorous, allegedly filthy and provoking disgust, men who refuse to marry open – that is, loose – girls incite to genital modification. 'Logic' dictates that the offensive parts must go.

Aware of this cultural encoding, contemporary activists and artists have revived the rose. Innumerable anti-FGM campaigns use the flower as their symbol, and so do engaged playwrights. In Stuttgart, for instance, on International Women's Day 2006, performance artist Dorothea Walter staged the underlying tension between wounding and recovery in a sketch titled "Love the Rose. On FGM."[8] The theme also inspired creative staff in European advertising agencies, answering a call for visuals by MEP (Member of the European Parliament) Alexander Alvaro. In DON'T. *Women's Art Protest against Female Genital Mutilation*, Katja Kamm adorns with contented female figures the uncut undulation of an upright bloom. The caption: "Let us worship our female nature … no more FGM" (11). "It's my nature. Don't destroy it" is likewise Elke Ehninger's plea (7).[9]

Not limited to Western portrayals, however, the two opposing themes, affection for the "(American) beauty" and prophylactic floral trimming, are equally captured in works of art protesting FGM by Nigerian artist Godfrey Williams-Okorodus.[10] In "Defiance of Pain 1" and "Defiance 2"[11] a golden rose dominates the canvas, yet the paintings differ as to the victims' options. In the first, sharing the visual center with a naked razor, the flower is grasped by a small girl, her back turned to a wall of women. Inexorable, they stand for custom, as not a chink appears in the barrier they form. Keeping her blossom is out of the question, despite the suggestion of "defiance" in the painting's title. Strikingly pale against the adults' deep hue, seated on the ground while grown-ups stay rooted in tradition, the young one is outnumbered and without allies. Can she rebel? Or is the artist alone insubordinate in visualizing the suppressed (Reinharz et al 12)?[12]

Godfrey Williams-Okorodus. *Defiance of Pain 1 and 2. Oil on canvas. 1998*

"Defiance 2" gives the girl a better chance, for here she dominates the foreground, arms crossed in rebelliousness. A flower next to her also glares in rich, red fullness, not yet faded like its companion bloom, while the village recedes behind both audacious blossom and evading teen. Will she escape? Possibly. But wait! There, to the left, is a palimpsest, a ghostly rendering of two imams, and a third hovering just off the margin, leading the story off of the canvas and back into the world. The men appear at once behind and in front of the girl. Blocked again.[13]

Roses suffuse still another Williams-Okorodus oil, "The Uhrobo Bride."[14] In the "Defiance" series, women are clearly associated with the genital attack, but the reason for female complicity goes unaddressed. The aptly titled "Bride" associates the amputations with enhancement of beauty and desire. Four elements stand out in the bust of a serene woman with harmonious sensual features: the same passionate crimson triangulates full lips, beaded necklace, and fat flower. The single discordant element, a razor adorns her right ear – the instrument aestheticized as jewelry. The same elision of disfigurement and fashion also frames the portrait: ten palms in an ambiguous gesture –stop? I give up?– beckon to the viewer, most tattooed with flora but one with a blade. Here Williams-Okorodus anticipates the "truth" most challenging to activists: women's emotional attachment to the damage, justified as pain in the service of beauty. As the French say, "Il faut souffrir pour être belle."

Is FGM then 'mere' cosmetic surgery? It is ethically unacceptable, of course, because forced on minors, but, analyzing "Female Circumcision among Nubians of Egypt," Fadwa El Guindi asks: "'Had *This* Been Your Face, Would You Leave It as Is?'"[15] Is it an aesthetic issue?

In large measure, yes, and for that reason resistant to arguments from health or rights. Motivated by idealism and their wounds, Maryam, Nura and Fadumo make valiant efforts to overcome the 'modesty' their upbringing taught them to value. They are indeed torn; for their emotions tell them that their suffering was not for naught, that the ordeal had done them good. Nura is shocked to find herself in Germany among a nation of the unclean. And Korn admits her loathing for genitals left in the raw.

"You stink," Fadumo, age ten, tells two Spanish sisters while their parents sip cocktails at her uncle's Mogadishu home. In this scene, Somali girls confront their European classmates, taunting them for the uncouth status of their nether parts, open, wet, or worse. "You're dirty, yuck!" Fadumo sneers. "You drool! You're going to hell!" As children will, the Spanish youngsters lie: "We're circumcised like you!" But Fadumo declares, "I don't believe it" and escalates doubt into assault. Once the girls have retreated to her bedroom, Fadumo challenges her guests. "Prove it! Show us!" When the Spanish girls don't, the Somalis do:

> Determined, I undid the button and unzipped my jeans, hesitated for a fraction of a second, then down went the pants over my bottom and I pulled aside -- not for long but long enough for everyone to see -- my panty crotch. One after the other, everyone did the same.

Everyone, that is, except the Spanish girls whose reluctance provokes attack:

> "Get 'em. Let's look for ourselves!" As if waiting for my command, the gang stormed, threw the sisters on the bed, ... and pinned their arms and legs while we removed their underpants. "I knew it," I cried. "Look at them! How wrinkled they are, how shriveled, how ugly. Yuck!"[16]

First, we find assailants who have themselves been seized, had their legs torn apart, and been constrained by a regiment of women. This, then, is learned behavior. Second, the anecdote conveys a point that no discussion of the topic should neglect. Not only is FGM *normal* to those whose ethnicity mandates it as a condition of gender identity and group membership, but beauty and cleanliness[17] are thought to result. These in turn give the girls pleasure – if they recover, that is, -- as they have longed for the event, to change status and become "women." Genital erasure is the sign of belonging – as only cut roses can adorn the home.

Hence, the scene reveals the practice's psychological appeal and its motor in peer pressure. To become a "positive deviant" then means making a sharp turn-around from pride to shame to an emotionally neutral knowledge from which activism may emerge. But this is a burdened mental move and takes *enormous* valor.

The needed audacity is not easily found. At Mt. Holyoke in 2004, encouraged by the screening of Ousmane Sembene's ground-breaking film *Moolaadé*, I spoke on "Somali Immigrant Memoirs and Campaigns to End FGM in Germany."[18] After the talk, two U.S.-born Somali students approached. The speech, they said, was "awesome." Would I give it again for the African and Caribbean Students Association? Of course, I replied, but on condition that we work together. So the three of us met several times to study and prepare presentations for one or two of the local prep schools. Thanksgiving, however, brought a change of heart. "We're really sorry," the girls said, "but our mothers have forbidden us to work on this." Why? If word should get back to the immigrant community, they would have been denounced as traitors.

A bouquet for Alice Walker[19]

Sensibilities can indeed be tender on this subject, as Alice Walker found out in the USA. In the early 1990s, Walker became the first personality of world renown to publicly oppose female genital mutilation. Her novel *Possessing the Secret of Joy* (1992) and documentary (with Pratibha Parmar) *Warrior Marks* (1993) did more than decades of international and grassroots agitation by 'ordinary' citizens to raise the issue with law- and policymakers in Africa and the African Diaspora. Before the 1990s, FGM could still be considered a taboo subject, the number of academic and popular articles remaining shamefully small. True, in 1982 Elizabeth Passmore Sanderson published an 82-page bibliography, but ethnographic work, the bulk of listings, addressed academics, not legislators, citizens, activists or adherents of the practice. And anthropologists, to put the best construction on it, although admirably aware of colonial abuses, tend to be blinded by good-will – toward systems in power, which means patriarchal privilege over women.

Walker, however, is not an anthropologist but a creative writer whose intervention was followed by action. I contend it is more than coincidence that the two major international conferences, Vienna on Human Rights (1993) and Cairo on population (1994), coming so closely on the heels of Walker's efforts, finally figured FGM as a major humanitarian challenge.

Yet, in the United States, where girls of African origin are clearly at risk, problems in reception of Walker's book and film surfaced immediately. Now, Walker knew to be fearful. In *Anything We Love Can Be Saved. A Writer's Activism*, she records her apprehension.

Visiting Jung's home in Bollingen, the final quaff of inspiration taken, she notes: "This was the last journey I had to make before beginning ... *Possessing the Secret of Joy*, a story whose subject frankly frightened me. An unpopular story. Even a taboo one."[20]

Although the USA gave the novel mixed reviews at best, the film incurred outright hostility. Led by African women intellectuals residing in North America, voices of resentment against Walker's violation of boundaries resonated loudly. *Newsweek*, for instance, quoted Sudan's premier female surgeon, Nahid Toubia, alleging that only because Walker's popularity had suffered did she take on FGM; a "falling star," the author was trying to get "the limelight" back.[21] More to the point, Walker, it was said, failed at empathy, her accusatory finger not illuminating but condemning and, therefore, insulting the audience she claimed to address, African women perpetrator-victims.

How affected Walker was by her critics can be teased out of the speech that opened *Warrior Marks'* tenth screening on February 24, 1994, in Oakland. "What can you do?" she asks her opponents. "... Refrain from spending more than ten minutes stoning or attempting to malign the messenger. Within those minutes thousands of children will be mutilated. Your idle words will have the rumble of muffled screams beneath them."[22] Yes, she concedes, victims "will have to stand up for themselves, and ... put an end to it. But that they need our help is indisputable."[23] Indeed.

Yet, ironically, Walker's compassion is siphoned from a pool of shared African-*American* suffering, "our centuries-long insecurity."[24] This presumption of solidarity, not with former slaves but with immigrants to the United States, proves the lightning rod to her African critics. She dares them to know "who we are [and] ... what we've done to ourselves in the name of religion, male domination, female shame or terrible ignorance."[25] But who are 'we'? An African-American assumes commonality with African immigrants in the USA – an unreciprocated move.

What is unbearable to Walker's critics but incontrovertibly central to events is named by Verena Stefan in a chapter devoted to Tashi in *Rauh, wild & frei: Mädchengestalten in der Literatur* [Rough, Tough and Free: Images of Girls in Literature]. Stefan reads the murder of the Tsunga (Walker's invented term for the exciseuse in *Possessing the Secret of Joy*) as a dagger to the shibboleth, reverence for the matriarch inculcated in part by an authoritarian culture.[26] Stefan reads the broken trust of FGM as "betrayal of girls by their mothers," and critiques the "control that the older wield over the younger."[27] Furthermore, in a revealing contrast, Stefan takes the rapport between Celie and Shug in *The Color Purple* as the other side of clitoridectomy's perfidy: "During circumcision, we have shared intimacy between woman and girl, but it is the intimacy of horror. Women observe and touch a young girl's genital – to mutilate it."[28]

In *The Dynamics of African Feminism. Defining and Classifying African Feminist Literatures,* Susan Arndt confronts homophobia, fear of which propels Stefan's observations, and also, I believe, accounts in part for Walker's negative U.S. reception. In particular, two theorists of African womanism – Mary E. Modupe Kolawole and Chikwenye Okonjo Ogunyemi – frankly come out as homophobes. Arndt writes, with inappropriate neutrality: "I do not know of any feminist theoretician in the West who dissociates him- or herself explicitly from lesbianism.... [So] it is a novelty within the feminist discourse that theoreticians of gender issues like Ogunyemi and Kolawole reject lesbian love explicitly, generally and firmly."[29]

Now, as most of us know, homophobia kills – and overcoming it is one of the last frontiers. It was homophobia that made 'offensive' to some critics *Possessing the Secret of Joy;* and the lesbian subtext is doubtless present in *Warrior Marks.* Pratibha Parmar, after all, lives an openly lesbian lifestyle. Although "a film maker first and last," and not a "lesbian filmmaker," Parmar has, for instance, in her film "'Jodie: An Icon' ... looked at ways in which ... Foster has been constructed ... for lesbians in her various screen personas."[30] Juxtaposed with a programmatic statement by novelist Buchi Emecheta, the conflict becomes clear. Emecheta charges Western feminists with concern only for "issues... relevant to themselves ... transplant[ed] onto Africa. Their own preoccupations – female sexuality, lesbianism and female circumcision – are not [African women's] priorities."[31]

I disagree. Consider Fanny Ann Eddy, a thirty-year-old human rights activist, who, in 2002, had founded a Lesbian and Gay Association in Sierra Leone. On September 29, 2004, Eddy was murdered, having earlier told the Human Rights Commission of the United

Nations in Geneva how dangerous it was for lesbians, gays, bi- and trans-sexuals to remain invisible in African society.[32]

Invisible, too, had been FGM. But breaking taboo has its price, and Walker suffered for her courage. Was she still ill at ease with the topic more than a decade later? Would her retirement be permanent? I asked Efua Dorkenoo, OBE, advisor and participant in *Warrior Marks*, if she knew why Alice had stopped campaigning. After all, I thought, Walker intervenes for other, important abuses and "insists upon the necessary and strong relationship between spirituality, activism and art."[33] Mainly, Efua told me, Alice had not forgotten the broad misrepresentation of her motives, the charges of ignorance of Africa and of self-interest. "That hurt her quite a lot, leading to the decision [to step back]. Her aim had always been to use her talent as an artist to bring the subject to the world. Having done that, maybe she feels justified in moving on to other things."[34]

Typical of critics who misunderstand her aims, Jo Ellen Fair sees Walker's presence in her documentary as "paternalism," giving it an aura of "Westerners know best."[35] But *Warrior Marks* is a mixed-genre, as much a portrait of the artist-activist as a plea for girls. The "Amt für Frauen" [Women's Buro] in Schöneberg, Berlin, agrees, praising the documentary's "poetic narrative." Like a *bildungs-roman*, "without recourse to bloody sensational scenes, [it] gets under your skin."[36] Behind Fair's judgment, in contrast, is an unconscious but tenacious a priori: only Africans own this issue. Awa Thiam, for one, firmly disagrees. When asked, "How do you feel about us coming here and making this film?" she tells Pratibha Parmar: "'You know, I work … in the belief [in] universal sisterhood, that we are all in this together."[37]

Sadly, where allies should be, they often aren't. Take, for instance, Yari Yari Pamberi, that remarkable gathering at NYU in October 2004, and Walker's presentation at the end of a star-studded panel. Not a word about FGM, this omission of a piece with Ousmane Sembene's *Moolaadé* having been on the conference schedule but taken off. Lincoln Center, it was alleged, provided a better venue. To his credit, as he announced the venue change, host and organizer Manthia Diawara gave a few militant words to FGM. But as recently as 2004, the elite of the African and African-American intelligentsia continued tight-lipped over torture.[38] Only now, a decade later and with the full financial and moral backing of both the *Guardian* newspaper and the British government's development agency DFiD has the situation begun to change.[39]

How different has been reception in Europe, particularly in Germany, where more than 20,000 African girls are at risk. Walker's input has not only been welcomed but acted upon. For instance, following a major interpellation in 1998, Walker is named by the Bundestag as an inspiration in asylum debates.[40] The decision to offer sanctuary to African women fleeing the threat of excision is, in part, based on Walker's work. "In many respects, genital mutilation resembles torture and violates the human right to bodily integrity," the Bundestag concurs.[41] And the document goes on: „Human rights organizations like Terre des Femmes and human rights activists like Alice Walker have been pushing the theme into public awareness for years." Footnoted is "Walker, Alice/Parmar, Pratibha, *Narben oder die Beschneidung der weiblichen Sexualität*, Hamburg 1996," the German translation of *Warrior Marks*.

What's more, at a preparatory multi-partisan hearing organized by the Green Party in Bonn, Dr. Angelika Köster-Lossack, MdB (Member of Parliament) embedded the German title of Walker's film – *Narben [Scars]* – in the title of her talk. Terres des Femmes, the German NGO advocating for women's human rights commended by the Bundestag, took its motto directly from Walker: "Resistance is the secret of joy," the closing lines of *Possessing the Secret of Joy*. Alice Schwarzer, Germany's 'first' feminist and editor of *EMMA*, has featured Walker and *Narben* in her magazine's pages. In 1996, Christa Müller, wife of Oskar Lafontaine, former head of the Social Democratic Party, advertised *Narben* on the program of her association's inaugural conference, the keynote address delivered by FORWARD'S Comfort Ottah who appears in Walker's film. In fact, *Warrior Marks* has been distributed throughout the nation; nearly every major university and gymnasium has it in its archives or has shown it at film festivals and other events.[42]

This astonishing diffusion and approval have not been due to suppression of resident African women's voices, for their opinions have been sought. They are engaged in many of the NGOs feeding into governmental agencies concerned with refugees, foreign aid, and women. Finally, government has accepted the responsibility of protection, even if there remains a great deal to be done toward implementation. Indeed, a network for all groups working on FGM within

Germany and abroad, INTEGRA,[43] held its inaugural conference in Berlin on December 12-13, 2006. It acts under the patronage of none other than the President of Germany, at that time Dr. Horst Köhler.

In conclusion, what has made such a difference in reception? Why should Alice Walker be shunned in the USA for dealing with this issue but celebrated in Europe, especially in Germany? Reasons are, of course, complex, and have to do with the development of the women's movement on both continents; on what has been learned, in Germany, from the Holocaust; on what is permitted in terms of who may speak 'for' whom, and on the preference for human rights discourse in Europe as opposed to post-structuralist shattering of needed coalitions in the USA with its stubborn, divisive, essentialist stance on "race." Though no one will argue the absence of racism in Europe, the history of its institutionalization is very, very different – and has led to a striking distinction in Alice Walker's influence on movements against FGM.

Returning to the Rose

Can it be a coincidence that Walker's most famous essay is titled "In Search of Our Mother's Gardens"? In it, she confronts the suppression of creativity in the formerly enslaved and their impoverished female descendants who, like Walker's mother, lacking canvas and oils, poured their souls into beauty afforded by nature – they cultivated gardens. Flora play an inordinate role in Walker's work. *The Color Purple* adorns the fields. The divine embodied in a wild flower, it dare not be overlooked, for "anything we love can be saved."

Happily, this promise even covers amputated organs, for a pioneer in reconstructive surgery can now restore the clitoris. Pierre Foldes, a urologist, has developed a procedure that consists of trimming the scar from the clitoral surface before rolling back tissue along the shaft to expose the still-existing organ, for the clitoris extends backward from its tip to the 3rd and 4th vertebrae. It cannot, therefore, be truly excised, and more than 75% of patients regain some sensitivity.

He reunites the cut rose with its root.

References

[1] Angier, Natalie. *Woman. An Intimate Geography.* NY: Anchor, 2000. 88.

[2] Abdi, Maryam Sheikh. "The Cut." 2006. http://www.popcouncil.org/rh/thecut.html Accessed 3 February 2007.

[3] Abdi, Nura and Leo G. Linder. *Tränen im Sand.* Bergisch Gladbach: Verlagsgruppe Lübbe, 2003. Excerpted as "Watering the Dunes with Tears." *In Feminist Europa. Review of Books.* 28-33. http://www.ddv-verlag.de/issn_1570_0038_jahr2003.4_vol01_nr01.pdf Rpt. *In Feminist Europa. Review of Books.* Special issue on FGM. http://www.ddv-verlag.de/issn_1570_0038_FE%2009_2010.pdf Retrieved 27 December 2014.

[4] Parrot and Cummings offer "Success Stories and Promising Practices" recording victim reduction in certain ethnic groups in Senegal, Uganda, Kenya and Burkina Faso as a result of targeted interventions or best practices. But the numbers saved remain (too) small. Parrot, Andrea and Nina Cummings. *Forsaken Females. The Global Brutalization of Women.* NY: Rowman & Littlefield, 2006. See also "Prevalence of Female Genital Mutilation by Country." *Wikipedia.* Web. http://en.wikipedia.org/wiki/Prevalence_of_female_genital_mutilation_by_country. Retrieved 8 January 2015. Based on a 2013 UNICEF Report, the latest news confirms that a diminution in some areas remains unacceptably small leaving thousands of girls still at risk.

[5] WHO. "Sexual and Reproduction Health. Female Genital Mutilation and Other harmful Traditional Practices." http://www.who.int/reproductivehealth/topics/fgm/prevalence/en/ Retrieved 27 December 2014.

[6] The least damaging intervention is clitoridectomy and yet Angier's words apply: "The unarguably vile practice goes by various names, including female genital mutilation, or FGM; African genital cutting; and female circumcision – although as many have pointed out, it is more akin to penile amputation than to male circumcision and should not be given the courtesy of comparison" (Angier, ibid. 86).

[7] Blake, William. "The Sick Rose." *Poetry Foundation* website. http://www.poetryfoundation.org/poem/172938 Retrieved 27 December 2014.

[8] Gruber, Franziska. "Aktionstag gegen Genitalverstümmelung." *Terres des Femmes – Menschenrechte für die Frau.* 2/2006. 12-13.

[9] Alvaro, Alexander, et al, eds. DON'T. *Women's Art Protest against Female Genital Mutilation.* Lünen, Germany: Drückerei Peter Holtkamp GmbH, 2006.

[10] "Through the Eyes of Nigerian Artists. Confronting Female Genital Mutilation" is an exhibition of paintings and sculpture that opened in 1998 in Nigeria, and was then shown at the Women in Africa and the African Diaspora (WAAD) conference at the University of Indiana-Purdue University. The artworks came to Germany in 2000, where 70 venues displayed them over the next six years. In the USA, the exhibition has been viewed at Brandeis, Harvard, Cornell, Bucknell, SUNY-Fredonia, Bridgewater State College and Monmouth University. It is available for rent from UnCUT/VOICES Press.

[11] "Female Genital Mutilation Through the Eyes of Nigerian Artists." FORWARD. Safeguarding rights & dignity. http://www.forwarduk.org.uk/key-issues/fgm/nigerian-artists Retrieved 27 December 2014.

[12] Reinharz, Shulamith, Tobe Levin and Joy Keshi Walker, eds. *Through the Eyes of Nigerian Artists. Confronting Female Genital Mutilation.* Frankfurt: Mainprint, 2006. 13.

[13] Reinharz et al, Ibid. 13.

[14] Ibid. 12.

[15] Abuscharaf, Rogaia Mustafa, ed. *Female Circumcision: Multi-cultural Perspectives*. Philadelphia: U. of Pennsylvania , 2006.

[16] Korn, Fadumo and Sabine Eichhorst. *Born in the Big Rains. A Memoir of Somalia and Survival.* Trans. and Afterword. Tobe Levin. NY: The Feminist Press, 2006. 119-121.

[17] Cleanliness not only as hygiene but also as (ritual) purity. See Mary Douglas on "how rituals create and control experience" (82). Douglas, Mary. *Purity and Danger. An analysis of concept [sic] of pollution and taboo.* NY: Routledge, 2005. (First published 1966).

[18] "Happenings October 2004." Mount Holyoke College. *College Street Journal.* https://www.mtholyoke.edu/offices/comm/csj/100104/index.shtml Retrieved 27 December 2014.

[19] For extended discussion of Alice Walker's influence on the movement against FGM see Tobe Levin, ed. *Waging Empathy.* Alice Walker, *Possessing the Secret of Joy, and the Global Movement to Ban FGM.* (Frankfurt am Main: UnCUT/VOICES Press, 2014); Tobe Levin. "Feminist (and "Womanist") as Public Intellectuals: Elfriede Jelinek and Alice Walker." *The New York Public Intellectuals and Beyond. Exploring Liberal Humanism, Jewish Identity, and the American Protest Tradition.* Eds. E. Goffman and D. Morris. (W. Lafayette, IN: Purdue U.P., 2009). 243-274.

[20] Alice Walker, "Heaven Belongs to You. *Warrior Marks* as a Liberation Film." *Anything We Love Can Be Saved. A Writer's Activism.* (NY: Random House, 1997). 147-151. Here 126.

[21] Tobe Levin, "Alice Walker, Matron of FORWARD." *Black Imagination and the Middle Passage.* Eds. Maria Diedrich, Henry Louis Gates, Jr. and Carl Pedersen. (NY: Oxford UP, 1999). 240-254. Here 244.

[22] Alice Walker, "Heaven Belongs to You" *Re-visioning Feminism around the World.* NY: The Feminist Press, 1995. 62-63. Here 63.

[23] Ibid.

[24] Walker, *Anything*, 150.

[25] Ibid.

[26] See Dympna Ugwu-Oju. *What will my Mother say? A tribal African girl comes of age in America.* Chicago: Bonus Books, 1995. Although African-American parents tend to be stricter than those of other U.S. minorities, their authoritarian child-rearing pales beside portrayals of parent-child interaction in Ugwu-Oju or in Sembene's *Moolaadé.*

[27] Stefan, Verena, *Rauh, wild & frei: Mädchengestalten in der Literatur* (Frankfurt/Main: Fischer, 1997). 108.

[28] Ibid.

[29] Arndt, Susan, *The Dynamics of African Feminism. Defining and Classifying African Feminist Literatures* (Trenton: Africa World Press, 2002), 53.

[30] Lolapress Europe. "Identities, Passions and Commitments. An interview with the British Filmmaker Pratibha Parmar." *Lola Press. International Feminist Magazine.* Nov. 99 – April 2000. No. 12. 36-41. Here 38.

[31] Ravell-Pinto, Thelma. „Buchi Emecheta at Spelman College." SAGE. *A Scholarly Journal on Black Women.* 2-1 (Spring 1985): 50-51. Here 50.

[32] Guido, Anke. "Menschenrechtsverletzungen an lesbischen Frauen." *TdF Menschenrechte für die Frau*. 2/2005. 12-13. Here 13.

[33] Griffin, Farah Jasmine. "The courage of her convictions." Rev. of Alice Walker. *Anything We Love Can Be Saved: A Writer's Activism*. NY: Random House, 1997, in *The Women's Review of Books*. Vol. XV, No. 4/ January 1998. 23-24. Here 23.

[34] Efua Dorkenoo. Telephone interview with the author. 18 April 2005.

[35] Jo Ellen Fair, "The Lives of Women in Africa." *Feminist Collections. A Quarterly of Women's Studies Resources*. Vol. 19. No. 4, Summer 1998. 11-13. Here 12.

[36] Amt für Frauen. Bezirksamt Schöneberg von Berlin. Flyer distributed at a showing of *Warrior Marks*. 28 May 1997. Rathaus Schöneberg.

[37] *Lolapress* Ibid. 36. Pratibha answers, "It was refreshing for me to hear her, having come from the late 80s and early 90s talk of post-feminism, with so much cynicism around the bankruptcy of feminism. It was good to meet a woman who still has that wonderfully optimistic belief in the idea of universal sisterhood, which many of us held in the 70s, but which got lost along the way as we got more fragmented and disparate" (*Lolapress* 36-37).

[38] Another distressing but typical example: Ada Uzuamaka Azodo, in „Issues in African Feminism: A Syllabus," includes nothing on FGM – in 1997! The course was offered at Indiana University Northwest.

[39] At the UN Commission of the Status of Women NGO conference in March 2013 Department for International Development (DFiD) chief Lynne Featherstone announced the UK government's allocation of 35 million British Pounds to fight FGM over the following five years. This is the highest sum ever earmarked in support of campaigns to stop the practice.

[40] See http://www.ak-kipro.de/fgm_bundestag200107.html Retrieved 17 July 2004.

[41] Antwort der Bundesregierung. Drucksache 14/5285. http://www.ak-kipro.de/fgm_bundestag200107.html Retrieved 5 April 2005.

[42] Levin "Alice" 246.

[43] To maintain his association's independence, Rüdiger Nehberg refused to enroll Target e.V. as a member of INTEGRA that now claims more than 30 members. Target, however, made headlines for a conference it sponsored on 22-23 November 2006 in Cairo, chaired by the Grand Mufti of Al Azhar University, Professor Ali Goma'a. Nehberg and his associate Annette Weber showed an explicit film of the operation to imams from throughout the Islamic world. The result was issuance of a Fatwa: "FGM is a crime that violates Islam's most cherished principles." (See Nehberg, Rüdiger and Annette Weber. *Karawane der Hoffnung*. http://www.ruediger-nehberg.de/ Accessed 1 December 2006).

[44] For more about Dr. Foldes and clitoral restoration see Prolongeau, Hubert. *Undoing FGM. Pierre Foldes, the Surgeon who Restores the Clitoris*. Foreword Bernard Kouchner. Trans. Tobe Levin. Frankfurt am Main: UnCUT/VOICES Press, 2011.

3

ROSE Symbolism in Qabalistic Tarot and Beyond

By Albert Amao

The Qabalistic[1] Tarot, a pictorial collection of seventy-eight Arcana or cards, is divided into twenty-two Major Arcana, the most symbolic keys of the Tarot system, and fifty-six Minor Arcana. This essay limits discussion to the twenty-two Major Arcana, numbered 0 to 21, and will survey the symbolism of the rose in them, explaining why the rose is an important symbol in the Rosicrucian Brotherhood. A second focus of the essay is the Qabalistic Tree of Life, a major component in metaphysical teachings.

The symbol of the rose appears seven times in the Major Arcana, which have often been considered archetypes to explore inner realms.[2] The rose's seven appearances in this system are not fortuitous. Seven has often been considered a sacred number in Holy Scriptures. The Christian *Bible* speaks of 'seven deadly sins' and the symbolic 'seven days of creation.' According to the book of Genesis, God created the universe in seven days. However, today it is accepted by many philosophers, historians and recently by major theologians, such as John Shelby Spong, a retired Episcopal Bishop of Newark New Jersey, that even the Christian *Bible* used symbolic writing at times. Spong states, "Belief in the historic accuracy of these texts no longer exists in academic circles..."[3] Although the Christian *Bible* recounts that the world was created in seven days, the story should not be translated literally. And yet, the Christian Bible is not the only place in which the number seven appears. In nature, the rainbow has seven colors. There are seven days in the week, and seven chakras or hidden energy centers within the human body. As the reoccurrence of the number seven in seemingly familiar aspects of daily life is not purely coincidental, neither are the seven appearances of the rose in the Major Arcana of the Tarot system.

The Qabalistic Tarot tradition is based on Qabalah, the esoteric interpretation of the Jewish Scriptures.[4] However, modern scholar Lewis Keizer, author of *The Esoteric Origins of Tarot: More than a*

Wicked Pack of Cards, refutes this claim, stating that the modern esoteric tarot decks are a development stemming from the medieval Italian Tarocchi, a pastime that was popular throughout Medieval Italy where the images were often interpreted in magical games.[5]

Throughout this essay, the term 'Tarot Keys' is used, instead of 'Tarot Cards,' in order to differentiate from the incorrect notion that the Tarot system is a means to foresee the future.[6] Each key of the Tarot system encapsulates a great deal of wisdom; thus the name Arcana, which the *American Heritage* Dictionary defines as "Specialized knowledge or detail that is mysterious to the average person."

There is a secret and subtle language in the symbols of the Tarot configuration which speaks directly to the subconscious mind. Few are aware of this arcane language because it is not traditionally taught on any level throughout the academy. Dr. Robert Wang, author of *The Qabalistic Tarot*, comments on the lack of focus that academic circles have given to the imagery of the Esoteric Tarot Keys, which he claims could be used as a tool for personal self-exploration. Wang explicates, "Despite the increased public interest, surprisingly little attention has been paid to the Tarot by academia, though the cards are a veritable gold mine of art history and metaphysical philosophy."[7] However, everyone is surrounded by symbols in myriad forms, as the symbols are part of daily life. In Freemasonry fraternities as well as metaphysical and religious organizations, symbols are presented as emblems of identification within institutions that share the same ideological or religious principles. Many of the symbols taken for granted are conventional signs of religious, mystical, or metaphysical organizations when in fact they are a subtle language to convey an invisible dimension that cannot be expressed in conventional words.[8] Dr. Cynthia Giles, Jungian Psychologist, explains the importance of symbols and consciousness-raising: "These multiple levels of the archetypes are all present in the Tarot images and it is for this reason that the Tarot works in a variety of ways. For the Tarot reader, images are the point where a process of archetypal analysis begins, and from the images unfold the behaviors and styles of consciousness."[9]

Over time, the symbols of the Tarot Keys have acquired layers of increasingly complex meaning. This evolution tells a great deal about perception of the inner world, the nature of life and of the universe. Furthermore, each Tarot Key constitutes a single entity that can unlock wisdom which already exists in the mind of the observer. The combination of one Tarot Key with another Key can bring to mind a wealth of knowledge for those who know how to properly view the images. That is why the sages of this wisdom, such as Paul Foster Case, perhaps one of the greatest Qabalists and Tarotists of the twentieth century, has admonished in his book, *The Tarot: A Key to the Wisdom of the Ages*, that a person cannot exhaust all the knowledge summarized in this Qabalah in a single lifetime.

Tarot key 0, The Fool (BOTA)

Explaining the symbolism of Tarot will begin with the first Tarot Key, numbered 0, often called The Fool. In this key, one can observe a young man at the top of a mountain, clutching a white rose in his left hand. Yet, he does so without pricking his fingers on the thorns. Since pricking one's fingers would generally cause a certain level of pain and suffering, the fact that the chap grasps the rose unaffected by the thorns would suggest freedom from any terrestrial affliction. The white rose signifies purity and innocence.[10] The general meaning of a rose is *desire;* in this case, white conveys the idea of innocence, purity and higher aspirations, giving the basic idea to this key which represents the Universal Life-force, the energy behind growth and development. This is reinforced by the gesture of the young man looking upward to the heavens, toward higher aspirations. Roses, along with the lily, can be observed in other Tarot Keys as well, such as the Key of the Magician.

In the Tarot Key numbered 1, a Magician stands in a garden of red roses and white lilies. Roses also hang above the Magician's head, insinuating a rudimentary shelter. In this key, the red rose contrasts with the white on the Fool's key. The meaning behind each hue is

Tarot Key 1, The Magician (BOTA)

different; the red rose signifies active desire. The white lilies in the garden indicate knowledge, higher aspirations, abstract truth, and purity of the soul. Moreover, the Magician is looking at the table, which represents the level of reality in which the Magician exits.[11] Displayed on the table are four items, a wand, a sword, a cup, and a coin. The wand symbolizes will, the sword represents action and ideas, the cup signifies imagination and emotions, and the coin is associated with material things. Each symbol suggests a task or element of daily life that one must master. Knowingly or unknowingly, every human being is a Magician by virtue of possessing the potential to create his own reality through the power of attention, thoughts and emotions. One would be a white magician with positive intentions and doing good deeds for neighbors. One would be a black magician if fostering negative emotions and trying to take advantage of others to satisfy egotistical needs.[12]

Contrary to Four Noble Truths of the Buddhist doctrine, which propose that humans should suppress desire to avoid suffering, Qabalistic doctrine considers desires to be blessings.[13] In Buddhism, desire for material possessions should be suppressed while desire for enlightenment should be fostered. In Qabalah, it is said that God leads humanity toward liberation *through* God's desires. In fact, desire is the motivational force that impels humanity to action. That is why masters of wisdom, such as Buddha himself, recommended that pupils cultivate a steady desire for enlightenment to allow the learner to endure all the effort and sacrifice necessary to reach the goal. A human being without desire is virtually dead with no reason to live purposefully. Essentially, the Magician represents craving for self-improvement. Like the Magician, humans have all the necessary tools before them, but it is a person's *will*, or desire, that drives the individual towards action.

In Tarot Key 1, something regarding the roses of the Magician's garden is worthy of a note. Roses have five petals, or some multiple of five; thus roses represent the Pentagram, the five-pointed star displayed on the Magician's table as well as the human being as the microcosm. In contrast, lilies have six petals and therefore stand for the hexagram, the macrocosm, that is the universe.[14] Putting together these two digits creates the algorithm sixty-five. And sixty-five is the numerical value of the Latin word LVX[15] (50+5+10 = 65), which translates to the word 'Light' in English. Furthermore, sixty-five is also the numerical value of the Hebrew word *Adonai*, which translates to the divine word 'Lord.'[16]

The next Tarot Key in which the rose is present is Key 3, The Empress. The fundamental idea underlining this Tarot Key is similar to the esoteric meaning behind the rose. The Empress typifies Venus, the desire that nature expressed through creative *imagination.* "She embodies creative processes and lively growth and the birth of the new."[17] This key is associated with the planet Venus. In the Rider/Waite version of the image, The Empress has the glyph of Venus on the shield which is on her right, and her entire garment reflects the Symbol of Venus.

Furthermore, looking closely at the Empress, one can see that she is pregnant. In fact her sovereignty as Empress derives from her reproductive power, the basis for the existence of life and for the flowing of the Life-force into new forms. This idea is reinforced by the existence of the stream and waterfall present behind the Empress. The stream symbolizes the water of life, and the waterfall suggests the idea of union of male and female. The male is symbolized by the stream flowing into the pool beneath it, which represents the female. This is a beautiful pictogram chosen to represent the reproductive

Key 3, The Empress (BOTA)

power of Mother Nature and the flowing of the Life-energy! The Empress in Tarot Key 3 represents Mother Nature. Without fecundity, we have no reproduction, and consequently, no life. The reproductive power of the Empress is demonstrated by her scepter, a symbol of power and sovereignty, and not coincidentally, the staff rests on her womb. The basic meaning of this key is therefore duplication, multiplication, and reproduction in all areas of life.[18] A scholar on Rosicrucianism, Qabalah and Tarot, Dr. Arthur Edward Waite has offered an explanation of this Tarot key by stating, "The card of the Empress signifies the door or gate by which an entrance is obtained into this life, as into the garden of Venus."[19] Moreover, on the right side of the Empress we see five red roses representing the five physical senses that incite desire; in other words, every desire is rooted in the the body. As we have seen, the rose represents the Pentagram, and the Pentagram is also a representation of humanity. As the principle of karma teaches, it can be said that humans were incarnated on earth by their own intense desire to come to this material plane in order to learn and expand their consciousness through physical experiences in life.[20]

The next key containing the rose is Key 5 in which two priests kneel in front of the High Priest, known as the Hierophant. The cleric on the left wears a robe decorated with roses; the gown of the priest on the right is adorned with lilies.[21] These same flowers are part of the magician's garden, thus standing for desire and knowledge. The Hierophant is the revealer and teacher of sacred mysteries. Only those ready to listen to their Inner Voice and who adopt a humble attitude will benefit from the mysteries revealed by the High Priest.[22]

Entitled 'The Lovers', Tarot Key 6, where the symbol of the rose is implicit, depicts the Christian Biblical Garden of Eden including Adam, Eve and the so-called Serpent of Temptation. Behind the woman is the Tree of Knowledge of Good and Evil, and behind the man is the Tree of Life.[23] The five fruits on the Tree of Knowledge of Good and Evil, the occult representation of the five red roses of the Magician's garden, stand for the five senses or tempters of man, which is why the Serpent of Temptation is coiling up on the tree behind Eve.

It is interesting to note that the Christian *Bible* never mentions the apple in the story of Adam and Eve despite what most people believe

Key 5, The Hierophant (BOTA)

Key 6, The Lovers

today. Even the *Catholic Encyclopedia* blames the apple for Eve's fatal mistake.[24] It should be noted, however, that when one cuts an apple horizontally down the middle, the Pentagram appears in each half, with one seed inside each of the figure's arms. Is it not marvelous? The Pentagram epitomizes humanity, the microcosm, and appears in the very tool of nature that allegedly caused The Fall. Continuing with the allegory of Adam and Eve, the Devil told Eve that the fruit was ripened and therefore edible. She was instructed that eating the fruit would open her eyes and make her like God, knowing good and evil.[25] Interestingly, the *esoteric* elucidation of this story reveals that the Tempter and the Redeemer are the same. At times, the devil is also called 'The Deceiver,' 'The Father of Lies,' or 'The Master of Appearances.' In fact, the devil deceives through humanity's own five physical senses and with the appearances presented in the world.[26]

The next key with roses is Key 8, Strength. The flowers are presented as a chain that encircles the waist of the lady and the neck of a lion that stands beside her. As previously mentioned, roses symbolize desire. Hence, the flower chain is a series of desires woven together. Rightly cultivated, desires are the most potent forms of suggestion. The young lady presented in this key represents the purified subconscious mind that has dominion over nature through creative imagination; she exerts control over the forces below the human level of manifestation, represented by the lion. Her white robe represents purity, wisdom, light, and knowledge.[27] The meaning behind this key is that the subconscious has mastered natural forces through the law of suggestion. When desires are purified, the subconscious, which has perfect power over the lion, will become an ally to humans instead of an enemy. In the Tarot doctrine, the lion, king of nature's animal kingdom, represents all subhuman forces of the cosmic energy. Thus, key 8 depicts the lion under the control of the lady clothed in white.[28]

The final representation of the rose in the Tarot's Major Arcana is in Key 13, Death. Demise, desolation, and darkness appear on first view, but occult awareness states that there is no death, only a process of transformation and change. The natural Law of Transformation brings about dissolution and change.[29] Careful observation of this key therefore reveals the symbol of a seed in the upper left corner and a white rose in the middle of the right side. The white rose is that of Key 0 and has the same meaning as the flower in the hand of the Fool, that is purity and innocence. Therefore, Death implies rebirth; this is insinuated by the fact that the yellow sun in the background of Key 13 is rising in the East. The river, an emblem of the flow of life, is making a bend directing its current to the sun. The seed in the upper left corner has five rays and symbolizes rebirth and life.[30]

Ultimately, this Tarot key symbolizes the ending of the Piscean Age, a time characterized by global religion, and the renaissance of humanity into a new era, the Aquarian Age, an age of people in an advanced stage of spiritual consciousness. This is insinuated in Tarot Key 13 by the fact that the key shows only one foot, indicating the end of the Piscean Age. In Astrology, signs have their correspondence to different parts of the human body and Pisces corresponds to the feet.[31] Along with Tarot, the rose has been associated with other spiritual circles and when related to religious imagery, the symbol acquires a special spiritual dimension. Meditation on it produces a profound effect in a person's psyche. An example is the Christian 'Rosy Cross.'[32]

Key 8, Strength

Key 13, Death (BOTA)

The 'Rosy Cross,' also known as 'Rose Cross,' is a familiar symbol of the Rosicrucians, an organization dedicated to the "systematic approach to the study of higher wisdom that empowers you to find the answers to your questions about the workings of the universe, the interconnectedness of all life, your higher purpose, and how it all fits together."[33] The Rosicrucian Brotherhood is a Christian sect founded in Europe in the fifteenth century according to some sources although others trace its origins to long before the birth of Christ in some part of North Africa. This disagreement notwithstanding, the brotherhood combined the emblem of the cross and the rose, and adopted this icon as the trademark of what began as a secret society.[34] In fact, the cross and the rose as separate emblems are two of the most ancient and universal symbols in human history. The original 'Rosy Cross' crystalized as a Christian symbol in the second century, was a red rose positioned in the center of a cross.[35] This living image has both a mystical and esoteric meaning.

In the nineteenth century, the symbol of the Rosy Cross was adopted by the Order of the Golden Dawn, an organization formed in 1888 that devotes itself to the study of ancient wisdom.[36] The magicians of the Golden Dawn transformed the Rosy Cross into a more sophisticated and complex icon as they incorporated Qabalistic, alchemical, and astrological symbols. When observed closely, the Golden Dawn Rosy Cross reveals a rosy-cross in its central point, representing the Absolute from which everything is manifested. The Golden Dawn Rose Cross expresses the meaning of the ineffable and unpronounceable word of the Tetragrammaton, the four letters used to re-

Varied Illustrations of the Rosy Cross

Symbol of a Cross   *Five Petals Rose*   *17th Century Rose Cross*

present the name of Yahweh, a name that must not be spoken aloud in the Jewish faith. Encircling the innermost point of the cross are three Hebrew letters, Aleph (א), Mem (ם), and Shin (ש), known as the mother letters in the Hebrew alphabet. In turn, the three mother letters are surrounded by the seven Hebrew double letters, and consecutively, the seven letters are encircled by the twelve simple letters to form a rose of twenty-two petals. The three mother letters correspond to the three outer planets, the seven double letters to the seven traditional planets,[37] and the twelve simple Hebrew letters correlate to the twelve zodiacal signs. In total, this makes twenty-two, equivalent to the twenty-two Tarot Keys and the twenty-two Cabalistic Paths of the Tree of Life.

The Golden Dawn Rosy Cross Lamen also embodies the Qabalistic Tree of Life, a mystical concept of Qabalah which is used to understand the nature of God. The middle Pillar of the Tree is represented by the vertical cross-bar, which has three squares and the rose composed of Hebrew letters or petals occupying the center of the cross. The yellow square at the top of the vertical line is the *Sefirah Kether*, Crown of the Tree of Life. The rose petals represent the Sefirah of Tifaret, or Beauty. The Sefirah Yesod, or Foundation, has a glyph of the hexagram surrounded by the planetary symbols and the sign of the sun at its center. Finally, the bottom square of the

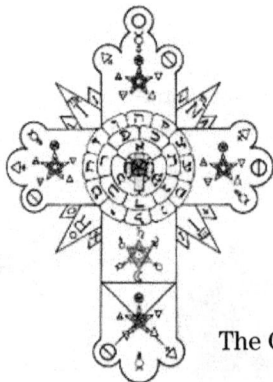

The Golden Dawn Rose Cross Lamen

vertical arm is the Sefirah Malkuth, or Kingdom. This is confirmed by the fact that this Sefirah is usually divided into four elements represented by the colors: citrine, olive, black and russet.

The other two columns of the Tree of Life are represented by the horizontal arms with four triangles adjacent to them standing for the Sefirot: Mercy, Severity, Victory and Splendor. These rays bear the Latin letters that form the acronym I.N.R.I., meaning 'Jesus of Nazareth, the King of the Jews.' Each letter is associated with an astrological sign according to its Qabalistic correspondence. For instance, the letter I, 'Yod' in Hebrew, corresponds to the astrological sign Virgo; the letter N, 'Nun,' to Scorpio; and the letter R, 'Resh,' to the planet Sun.[38]

Another representation of the Golden Dawn Rose Cross Lamen depicts each arm of the cross with an upright pentagram containing the symbol of the four alchemical elements, Fire, Water, Air and Earth, and crowned with the symbol of the Spirit at the top of the pentagram. As mentioned, the pentagram, in esoteric geometry, stands for the dominion of the Spirit over the Elements. At the end of each arm of the cross we find three alchemical symbols representing sulfur, salt, and mercury.

In Sacred Geometry, the cross is defined as the meeting of two lines, one horizontal and the other vertical, suggesting two perpendicular forces joined at a central point. The horizontal line represents the feminine and the vertical line the masculine principle. The icon also represents the physical world, epitomizing the dual nature of physical manifestation, that is, time and space, the vertical line standing for time and the horizontal one for space. Their intersection at a central point is an initial expression of the Life-force.[39] Incidentally, in Egyptian mythology, the crux Ansata resembles the Venus glyph, or symbol, ♀. In astrology, this means life, and as mentioned before, it appeared in the Tarot Key of the pregnant Empress.

The Rosy Cross symbol is sometimes called a Calvary or Piscean Cross because the vertical cross-bar is longer than the horizontal one, characteristic of the cross during the Piscean Age. A rose is located over the intersection of the cross and encircled by eight thorns in the form of five-pointed stars. The cross is also surrounded by 'Ouroboros,' which is a serpent swallowing its tail. The 'Ouroboros' is an ancient universal symbol depicting time, continuity of life, wisdom, and eternity.[40] The tip of the serpent's tail equates to a point, and its head is a circle; hence, together, they make the astrological glyph of the sun, ☉, that is, the point inside a circle. In the triangle over the cross we find the inscription of the Divine Name, Yod-Heh-Vav-Heh, in Hebrew characters, יהוה. As indicated above, one eso-

The Rose over a Calvary Cross

teric meaning of the cross is the material plane; the circular serpent depicts the cycle of death and rebirth. Here the rose epitomizes the unfolding of spiritual nature from the material one through pain and suffering.

Along with the cross, the symbol of the rose also has a deep mystical history that would later translate to religious significance as well. As mentioned, white roses symbolize purity, desire, aspiration, virginity, truth, and wisdom while red ones stand for passion, fertility, life, death, and, for a Christian analogy, the blood of Jesus Christ as well as the martyrs who died in his name.[41] The Rose Cross illustrates the triumph of spirit over matter, where the Rose, a symbol for love, is placed over the Cross, matter. The hidden significance reveals the redemptive power of Jesus Christ's blood shed on the cross when, according to Christian tradition, he was crucified on the hill of Calvary;[42] therefore, the message of the Christian Rosy Cross is redemption through a Savior.

As you can see, the rose symbol is not coincidental in the Tarot tradition. Significant meaning stands behind each pictogram, especially the rose and its appearances in the Major Arcana. The significance behind such an emblem is carried through other esoteric circles, including the Rosicrucians, where the rose was adopted as their official emblem. Both the rose and its meaning cannot be overlooked when considering esoteric doctrine, for the symbol of this life force is present in everyone's daily life.

Note: The BOTA Tarot keys have been reprinted with permission of the Builders of the Adytum, 5101 North Figueroa St., Los Angeles, CA 90042. (www.BOTA.org). Disclaimer: "Permission to use Builders of the Adytum images in no way constitutes endorsement of the material presented in this book."

Note: The author has made every effort to give credit for images of the rosy cross used for educational purposes here. The author and the editors cannot, therefore, be held liable.

Many thanks to Lisa Cucciniello for graciously editing this essay.

References

[1] The term 'Qabalah' is perhaps most recognized when spelled 'Kaballah.' However, in ancient Hebrew there was no letter 'K'. Therefore, for the purpose of this essay, the ancient spelling of this teaching will be used. This essay is an adapted and expanded version of a subchapter of my book entitled *Aquarian Age and the Andean Prophecy* (Indiana: AuthorHouse, 2006).

[2] See Paul A. Clark, *TAROT: The Magical Keys to Consciousness.* Paul Clark is the founder of the metaphysical school: Fraternity of the Hidden Light.

[3] John Shelby Spong, *BORN OF A WOMAN: A Bishop Rethinks the Virgin Birth and the Treatment of Women by a Male-Dominated Church* (San Francisco: Harper Collins Publishers, 1992) 3.

[4] See *The Columbia Electronic Encyclopedia*, Sixth Edition, (Columbia University Press, 2003).

[5] For further information on the origins of the Esoteric Tarot see *The Esoteric Origins of Tarot: More than a Wicked Pack of Cards* by Lewis Keizer, PhD, and Chapter II of *Beyond Conventional Wisdom.*

[6] Refer to Albert Amao's *Beyond Conventional Wisdom* (Indiana: AuthorHouse 2006).

[7] Robert Wang, *The Qabalistic Tarot* (Samuel Weiser, Inc., 1983) 2.

[8] For more information about the special language of symbols refer to the third chapter of the book *Aquarian Age and the Andean Prophecy* mentioned above.

[9] Cynthia Giles, *The Tarot, History, Mystery, and Lore.* (NP: A Fireside Book, 1994) 60.

[10] Hajo Banzhaf, *Tarot for Everyone* (Urania, Switzerland: AGM AGMuller, 2005) 32. See also Anthony Louis, *Tarot: Plain and Simple* (St. Paul Minnesota: Llewellyn Publications, 2004) 52.

[11] Ibid. 34.

[12] Rev. Ann Davies, *Inspirational Thoughts on the Tarot* (Builders of the Adytum) 8.

[13] According to Buddhist teaching, the Four Noble Truths are 1. Suffering exists; 2. Suffering arises from attachment to desires; 3. Suffering ceases when attachment to desire ceases; 4. Freedom from suffering is possible by practicing the Eightfold Path. For more information regarding the Four Noble Truths, see: "Four Noble Truths." *Britannica Concise Encyclopedia* (np: Encyclopedia Britannica, Inc., 2006).

[14] Paul F. Case, *The Tarot, A Key to the Wisdom of the Ages.* (np: BOTA, 1990) 45. See also Éliphas Levi, *Transcendental Magic, Its Doctrine & Ritual* (London: Braken Books, 1995) 89-98.

[15] Classical Latin had no letter 'U', so the letter 'V' would be in place of the letter 'U'.

[16] Israel Regardie, *Foundations of Practical Magic* (np: The Aquarian Press, 1979) 127.

[17] Hajo Banzhaf, *Tarot for Everyone.* 38.

[18] Ibid.

[19] Dr. Arthur Edward Waite, *The Pictorial key to the Tarot* (NP: U.S. Games System, Inc, 1989) 83.

[20] See *A Dictionary of Buddhism* (Oxford: Oxford UP, 2004).

[21] To fully appreciate the color pictures, it is recommended that you procure the book of Paul F. Case, *The Tarot: A Key to the Wisdom of the Ages.*

[22] Hajo Banzhaf, *Tarot for Everyone.* 42.

[23] Ibid. 44.

[24] See *Catholic Encyclopedia*, s.v. 'Sext.'

[25] Genesis, Chapter 3.

[26] This assertion has been extensively discussed in the essay, "The Adversary and the Redeemer," which can be found in the book *Beyond Conventional Wisdom* by Albert Amao.

[27] Banzhaf, Ibid. 54.

[28] Ibid.

[29] Refer to Albert Amao, "The Dialectic and the Hermetic Philosophy," *Beyond Conventional Wisdom*.

[30] Banzhaf. Ibid. 58.

[31] For further information refer to *Aquarian Age & the Andean Prophecy* by Albert Amao.

[32] For a more complete diagram as well as further links to the meaning of the 'Rosy Cross' and the emblems shown, consult http://altreligion.about.com/library/glossary/symbols/bldefsrosecross.htm. Retrieved 21 January 2015.

[33] For more information on the Rosicrucian organization, visit their official website, http://www.rosicrucian.org/home.html. Retrieved 21 January 2015.

[34] Mark O'Connell and Raje Airey. *Signs and Symbols* (Hermes House), 175. Also see *Catholic Encyclopedia*, s.v. 'Rosicrucians.'

[35] Ibid.108.

[36] See "Order of the Golden Dawn," *The Concise Oxford Companion to Irish Literature* (Oxford: Oxford University Press, 2003).

[37] For convenience of this essay, the Moon and the Sun are considered as planets.

[38] For more information on Kabbalah, visit the Kabbalah Centre's official website, http://www.kabbalah.com.

[39] Ibid.

[40] Mark O'Connell and Raje Airey, *Signs and Symbols* (Hermes House), 236.

[41] Ibid. 175.

[42] See *Catholic Encyclopedia*, s.v. 'crucifixion.'

4

The Tradition of the Rose in East Indian Culture and Ayurvedic Medicine

By Monika Joshi

Roses have an enduring presence throughout East Indian cultural history. The flower was used and is still used today in ceremonies such as Hindu weddings and in the practice of Ayurvedic medicine, a traditional system of Indian healing.[1] Applications of Ayurveda make extraordinary use of the blossoms, such as in preparation of rose fragrances. Often in garlands, these floral decorations serve as a means of personal adornment. As a basic ingredient of perfumes, the incense industry makes use of the flower's derivatives. Rose petals find their way into culinary preparations, like in *The Great Curries of India*.[2] The purpose of this essay is to familiarize readers with the historical, religious and medical significance of roses in East Indian culture and Ayurvedic medicine.

Historical records about the varieties of native Indian roses are very limited, except for ancient texts of Ayurveda, which identify the rose as a popular medicinal ingredient. However, within the last thirty years, archaeological discoveries indicate that the flower was already in use in the early Indus Valley civilization as one of the basic ingredients for preparing aromatic oils. In 1975, Dr. Paolo Rovesti, director of the Instituto Derivati Vegetali in Milan, led an expedition to Pakistan, an independent country that was part of India until 1947, to investigate the early Indus Valley civilization. During his expedition, he found a perfectly preserved terracotta distillation apparatus in the Museum of Taxila, dating from about 3000 B.C. The presence of perfume containers exhibited in the museum, dating from the same period, confirms the preparation of aromatic oils in the Indus Valley civilization.[3] However, it was the fifteenth century large scale introduction and promotion that led to the popularity of roses throughout India. The credit for this promotion is usually given to the Mughal dynasty, a Muslim ruling family founded by Babur which held power in India from the early sixteenth to the eighteenth century.[4]

Babur was a descendant of Genghis Khan, a thirteenth century ruler who founded an empire that included parts of China, Central Asia, the Middle East and Europe. In 1525, Babur established himself as sultan by marching his army from Kabul, Afghanistan, to Punjab, a region of the northwestern Indian subcontinent, and conquering all of India. Following Babur, the Mughal Empire produced five other great emperors, Humayun, Akbar, Jahangir, Shahjahan and Aurangjeb, who were collectively responsible for creating a massive empire in India. It was during the Mughal period that India saw a golden age of arts, military advancement, architecture and gardens.[5] Emperor Akbar, and later his son Jahangir, are credited with developing the Mughal style of art in India. Portraiture became prevalent in the time of Akbar, and he immortalized the rose through a portrait of himself holding the flower in his hand. Mughal art developed further during the time of Jahangir, and his fascination with natural science resulted in paintings of birds, flowers and animals.[6] One of the most famous literary and art works that features a rose is "Gulistan, Garden of Sadi," a poetic manuscript commissioned by Shahjahan that contains paintings of roses.

Subsequent to the Mughals, the East India Company is credited with introducing a wide variety of roses into India. The East India Company was a commercial trading venture founded by the British in the 1700s. Eventually it would rule India as it acquired governmental and military functions. Roy E. Sheperd, a historian and author of "History of the Rose," an article for *Science Education*, writes that some species of roses have been cultivated by the East India Company in the botanical gardens of Calcutta.[7] Today, roses grace many of the gardens found throughout India.

Featured prominently in Shalimar Garden of Srinagar, the capital of Kashmir in the northernmost tip of India, a garden of roses is a legacy of Mughal emperor Shahjahan. This site demonstrates beautiful contours through the channeling of a stream of water that flows down to the northeast corner of the lake from the hills above. Originally, the garden could only be accessed by water, but it can now be reached by a road that runs around the Dal Lake.[8] Coming south from Kashmir, tourists will find *Zakir Gulab Bagh*, a famous rose garden in the area of Chandigarh. Dr. M.S. Randhawa, the first chief minister of Chandigarh and a patron of horticulture, is the mastermind planner of this plot. Established in 1967, the thirty acre garden is named after India's former president Dr. Zakir Hussain and includes interplay of medicinal trees, fountains, and an array of roses in different colors, shapes and sizes. This garden, with its ornately laid-out lawns and flower beds, has 1600 different species of roses which include both natural and hybrid varieties. Thousands of people visit its annual rose festival in February and March.[9]

Further south in Delhi, the capital of India, are the skillfully planned Mughal Gardens surrounding *Rashtrapati Bhavan*, the official presidential residence, which are open to the public February and March of every year. The gardens are laid out on 130 acres and have 250 different varieties of roses. The Circular Garden area has 128 varieties of roses such as 'Tajmahal' and 'Arjun.' The 'Oklahoma' variety has been described by a frequent visitor as 'the darkest maroon rose', nearly black.[10] The landscape also features blue roses called 'Paradise' and 'Blue Moon' as well as a rare variety of green roses.

Bangalore, a city in south India, serves as an industrial and educational center for the country. It is also known as the 'Garden City of India' in honor of its beautiful gardens. Indian Muslim ruler Hyder Ali and later his son Tipu Sultan commissioned a large botanical garden in the eighteenth century. Ali decided to create this garden, called 'Lalbagh,' as a means of rivaling the Mughal Gardens that were gaining popularity at that time. *Lal*, in Hindi, the national language of India, stands for red, and Bagh stands for garden. It is said that Muslim emperor Tipu Sultan exclaimed, *"Lal Bagh!"* when he saw the garden's profusion of red roses along with its aesthetically designed lawns, flowerbeds, lotus pools and fountains. It is laid out on 240 acres and has hundreds of varieties of roses. It hosts annual flower shows which draw thousands of visitors.[11] Along with promoting visual pleasure in the botanical gardens previously mentioned, roses are used in their natural form in India's elaborate wedding ceremonies that are popular throughout the country.

These lavish occasions provide a great market for fresh flowers, especially roses. The use of the flowers begins with the decoration of the floral archway that leads to the wedding hall. These arches are covered with flower petals arranged to form beautiful patterns, such as the names of the bride and groom. During the ceremony, couples often exchange rose garlands, and the wedding *mandap*, a specially prepared stage or enclosure, is often decorated with roses.

This structure contains support beams from which strings of flowers and mango leaves are hung to form an elaborate canopy where the wedding takes place. After the vows, a grand reception is held where the entire background of the stage for the bride and groom is decorated with countless strings of flowers. The wedding night also makes abundant use of the roses to embellish the nuptial bed. Canopied beds are specially prepared with flowers, the most preferred being roses in order to create a curtain hung from all four posts. Rose petals are strewn on the marital mattress to create an inviting and memorable setting for the young couple on their first night together. Along with wedding rituals, roses are widely used in religious practices.

In worship ceremonies, garlands of roses are frequently offered to deities. Such an offering is the favorite of the Hindu God Krishna. In his book *Sacred Waters*, western author Stephen Alter recounts personal experiences of using roses in worship. On a pilgrimage to the Ganges River, the most sacred river in India, a priest placed flower petals, jasmine, rose and marigold, in his cupped palms and told him to release them into the Holy Ganges. He also recounts that he was instructed to sprinkle rose petals on the statue of Krishna.[12] During a visit to *Vrindavan*, 135 kilometers south of Delhi in the state of Uttar Pradesh, a place renowned for its temples of Lord Krishna, an eyewitness saw the roads leading to the temples filled with vendors on both sides. The rose vendors were ready with beautiful, fragrant roses before devotees started pouring in for morning prayers. The scent of the roses was so intoxicating that it stayed with the visitor on the long drive back to Delhi.[13]

According to the Brahma Kumari World Spiritual organization, which originated in India, roses are symbolic of surrender. The falling of each rose petal, one by one, is compared to the human surrender to God in all aspects of life. Shree Kripaluji Maharaj, founder of *Jagadguru Kripalu Parishat*, spiritual master and a poet-saint of recent times, has influenced thousands of devotees around the world in the *Bhakti Marg*, the path of love or devotion to God.[14] He calls the rose "King of Flowers" for its beauty and fragrance. Along with religious ceremonies, the rose also has a place in Indian folklore.

While the lotus finds mention in the early Vedic texts of ancient India, roses are rarely featured with the exception of the legend of

Padmavati, dated much later. According to an ancient Hindu legend, the goddess *Lakshmi* is said to have been created from 108 large and 1008 small rose petals. The story relates that both God *Brahma* and God *Vishnu* had an argument as to which was more beautiful, the lotus or the rose.[15] While Vishnu said it was the rose, Brahma, who had never seen a rose, refused to accept it. Finally, when Brahma saw a rose, he was so enamored with the beauty of the flower that he created the goddess Lakshmi out of its petals.[16] As mentioned earlier, the traditional form of Ayurvedic Indian healing is steeped in the tradition of the rose.

The language of Ayurveda is Sanskrit, one of the oldest written tongues. The roots of the word are *Ayus* and *Vedas*. *Ayus* may be translated as 'life' and *Vedas* as 'knowledge.' Hence, Ayurveda is the 'Knowledge of Life.'[17] According to astronomical records in the ancient Vedas, the most authoritative Hindu sacred texts, Ayurveda was in practice before 4000 B.C. Ayurveda teaches that man and woman are microcosms, a universe within him- and herself. Each is a child of the cosmic forces of the external environment, the macrocosm, and hence each individual existence is indivisible from the total cosmic manifestation.[18] Ayurveda views health and "dis-ease" in holistic terms, taking into consideration the inherent relationship between the individual and the cosmic spirit, individual and cosmic consciousness, and energy and matter.[19] The wisdom of Ayurveda was intuitively received in the hearts of rishis, the seers of truth. They perceived that consciousness was energy manifested into the five basic principles, or elements; Ether or space, Air, Water, Fire and Earth. This concept of the five elements is the heart of Ayurvedic science, which states that the five basic elements present in all matter also exist within each individual and manifest themselves in the functioning of the five senses of hearing, touch, vision, taste and smell. The physiology of the human being is governed by three main forces which are commonly known as the three doshas or humors, called Vata, Pitta and Kapha. Vata is made up of air and ether, Pitta is made up of fire and water and Kapha is made up of water and earth. These three humors govern all the biological, psychological and physiological functions of the body, mind and consciousness. If they are out of balance, the result is disease.[20]

Ayurveda recognizes the unique physical and mental constitution of an individual and detects the current state of imbalance through various techniques. It then offers a wide range of modalities to enable individuals to take care of their health. Ayurveda is not a system of healing in which everyone enjoys the same practices, and rarely are two treatment plans alike. Healing through Ayurveda involves all five senses, and the rose is used in many of these treatments. The sense of taste is addressed through proper diet and herbs, and roses are often combined with herbs which are then administered as an integral part of the healing process. Aromatherapy addresses the sense of smell, of which rose fragrances are often present. Vision is addressed through color therapy, and roses can be utilized for this purpose. Music and sound energies apply to the sense of hearing. Finally, the sense of touch is administered to through specially prepared herbal oil massages, of which rose oil is often an ingredient.

The cooling and soothing qualities of roses are used in various treatments. Rose petals, with their sweet taste and cool properties, have an excellent post-digestive effect. Roses are also used in laxative and digestive formulas because they enhance absorption. Emotionally, roses help increase feelings of love, compassion and devotion. Rose derivatives, like rose petal powder, massage oil, or aromatherapy oil, are alterative, antimicrobial and antipyretic. They cleanse the eyes and nourish the female reproductive system.[21] Cold poultices with rose and sandalwood pastes are sometimes prescribed to alleviate any kind of burning sensations. When used over time, the "essence of rose" pacifies heat in the blood, thus keeping imbalances in the skin from erupting. This is often used in acne treatments in order to remove any dullness in the skin. Rose derivatives are unsurpassed beauty oils. Rose-based oils benefit every skin type and are often used in lotions to treat infected, dry, or sensitive skin. Conjunctivitis, eye pain, irritation and various inflammations of the eyes can be treated with rose water.

Aromatherapy is an integral part of Ayurveda. Unlike foods or herbs, aromas work more directly on the mind and body and are often an important part of treating psychological disorders.[22] According to Ayurvedic texts, roses are connected with physiological factors that govern emotions and their effects on the heart. Rose extracts are soothing to both the heart and its emotions. The fragrance of rose oil produces a gentle but effective antidepressant effect. It brings joy to the heart, promotes feelings of love, reduces fear, drives away melancholy, helps recovery from grief and also settles the heart by combating the unwillingness to let go.[23] As a perfume, rose fragrances represent the essence of purity and innocence, yet can also act as an aphrodisiac. *Gulkand,* an Ayurvedic jam made primarily of rose petals and rejuvenating herbs, is used as a cooling tonic to combat fatigue and heat-related conditions; it is also naturally rich in calcium and has antioxidant properties.[24]

The ancient science of Ayurveda and its rose-focused prescriptions continue to be used effectively in modern days and can be applied anywhere in the world. One example occurred in Fresno, California, on 18 January 2007. A terminally ill eighty-seven year old hospital patient had liver cancer and was given two weeks to live. The family requested an Ayurvedic practitioner to help the man through his transition from this life. Rose water was sprayed around the patient and soothing chants were recited in the background. This rose therapy helped the man to be calm and therefore well rested, and even helped his roommate sleep through the night for the first time. The nurses were amazed, but it is believed that the high frequency effects of the rose water and chants created this transformation of atmosphere in the room and ultimately in the patients.[25]

Indian culture has a strong emphasis on religion and spirituality and gives much importance to the role that aromas can play in evoking divine inspiration. It is believed that the aroma of sandalwood, jasmine or rose can carry the human spirit to sublime levels. Rich aromas of plants, herbs and flowers have been tapped and incorporated as pure essences, often in oils, to create an atmosphere conducive to spirituality. In addition, these fragrances, in the form of anointing liquids, whether oil or water based, are used in personal adornment. Perhaps one of the oldest texts that record the use of fragrant oils for this purpose is the *Kamasutra,* the comprehensive text on sex and sexuality attributed to the sage *Vatsyayana* in the early fourth century A.D. The use of perfumed oils is listed as one of the sixty-four arts of which women should possess great knowledge. *Vatsyayana* also counsels that an ideal man would apply a limited quantity of ointments and perfumes to his body.[26] In another chapter, the sage talks about the use of fresh flowers and ointments as a means to create the atmosphere for sexual congress. Reference to "the pleasure-room, decorated with flowers, and fragrant with perfumes" appears in the voluminous treatise on sexuality.[27]

An Indian legend attributes the discovery of rose oils to the Mughal emperor Jahangir in 1612. According to the legend, the palace gardens of the Emperor Jahangir were interspersed with canals and fountains. Rose petals were strewn on the water to form a picturesque spectacle for special occasions. The pools and ponds of his estate were decorated with rose petals in preparation for his wedding at which point a film of oil was seen floating on the water. One version of the legend says that it was Jahangir who noticed the oily film, and another version credits his betrothed. The royal personages, so enamored with the scent of the oil, ordered that the oil be bottled for later use; thus began dissemination of rose oil perfume. However, the discovery of the ancient distillation apparatus mentioned earlier proves that rose oil was used much earlier than this seventeenth century legend.

Today, production of high quality rose oil is labor intensive. Compared to other plants such as lavender and rosemary, rose petals secrete an infinitesimal amount of essential oil. It takes 3,500 kilos of flowers to produce one kilo of oil. This amounts to 1,400,000 handpicked blossoms to produce thirty five kilos of oil, 40,000 blossoms for one ounce of oil, and sixty-seven blossoms to make one drop of oil. Harvesting must take place in the early morning hours before the heat of the sun evaporates the oil.[28] Roses, like all flowers, have a bio-rhythm that dictates cycles of blossoming and production of fragrance molecules. The harvesting of roses for distillation begins in the early dawn at a time the Vedas call *Brahma Muhurta*, 'God's time.' At these pre-dawn hours the rose petals are at the peak production of their oils. Roses produce their maximum levels of damascenone, the primary molecule of rose fragrance, on the mornings of a full moon.[29]

According to David Crow, author of *Floracopia*, more than 5,000 varieties of roses exist, yet only a few give the fragrance sought by perfumeries, the most popular being pink to light red 'Damascus' roses. A rose specialty 'Attar of Rose' is a traditional perfume of India that is composed of essential oils distilled into a base of sandalwood oil. The process of distilling roses for attar is rather simple; roses are harvested in April and May and placed in large copper vessels which are connected to the receiving vessel with a bamboo pipe that is covered with special grass rope for insulation. Flowers are placed in water-filled distilling vessels and are then heated. The vapors pass to a receiving vessel containing sandalwood oil, which absorbs the rose essence over a period of fifteen to twenty days, with new flowers being distilled every day. Rose attar, with its rich fragrant and intoxicating aroma, was often used as an aphrodisiac in ancient perfume aromatics and anointing oils.[30]

Today there is a concerted effort by the Indian government to step up the production of roses, particularly for the fragrance industry. It is being promoted heavily by government research institutes, such as the Institute of Himalayan Bio Resource Technology, which is engaged not only in the cultivation but also in the development of better and more affordable distilling processes.[31] The prominence given to this flower by leaders and celebrities has ensured that the rose continues to hold a special place in the country. The most famous personality of India, the first prime minister, Pandit Jawaharlal Nehru, always sported a rose in his buttonhole.[32] Movie personalities, politicians and other celebrities are often honored with garlands of roses from fans striving to outdo each other in the quantity of the flowers. A famous South Indian movie actor, Rajanikanth, made headlines on his forty-ninth birthday in December of 1998, when he was presented with a garland of red roses weighing 25 kilograms.

A view of the culture of India would not be complete without mention of the Indian movie songs that make ample use of roses in their lyrics. *Gulaabi aankhen jo teri dekhi sharaabi yeh dil ho gaya* are lyrics of a song sung by famed playback singer Mohammed Rafi in the movie *The Train*. The phrase translates to 'Looking at your beautiful rose-like eyes, my heart is drunk with your beauty.'[33] Another song from the film *Mantra* sung by Faakir is *Aankhon main shabab hai, chehra yeh gulab hai* which means, 'Your eyes have a magic and your face is like a rose.'

There is an interesting blend of diversity, rich culture, religion and tradition that permeates every aspect of life in India – with the rose being a significant part. In fact, India has given a lasting tribute to the beauty and fragrance of roses by unveiling scented stamps titled *Fragrance of Roses* to commemorate Valentine's Day in February 2007. One must consider the following scenarios of daily life in India.

In Delhi, the capital city of India, crowded bazaars of small shops are placed side by side flanking the narrow roads leading to the temples. Baskets of colorful roses vie with garlands of jasmine and marigold. Women from all areas sport jasmines, marigolds or roses in their well-oiled, long black hair. In the center of the most holy inner court of the temple is the statue of a deity covered in garlands of flowers. Priests chant sacred hymns while showering petals onto the god's likeness. The choice flower is the rose, premier among all other floral species. After worship, the priest distributes sacred water, vermillion powder, and petals or whole flowers to all in attendance. Women graciously accept the flowers, vying for the roses which they carefully tuck into their hair on their way out of the temple. In one more scenario, crowds gather around a priest on the banks of the River Ganges. The priest solemnly chants *mantras*, sacred words which are said to have originated from the Hindu text, the *Vedas*.[34] Hindus believe that these words have powers when uttered with the right intonation and bring many blessings to them when used as part of any ritual. The priests also have specific gestures that accompany the assemblage to strew the petals they are holding in their cupped hands into the river. Upon being released, the red rose petals float gently on the surface as they are carried away by the current.

Roses play an important role in Indian culture and will continue adorning buttonholes, peeping out from black tresses, and adding their own color to the vibrant sub-continent. But more than just a personal ornament, the rose is valued for its healing properties, such as in Ayurvedic treatments, and religious traditions, such as adorning statues of Krishna. The rose, with all of its properties, permeates the daily culture of East Indian life.

References

[1] David Frawley. *Ayurvedic Healing: A Comprehensive Guide*. (Delhi: Motilal Banarsidass Private Limited, 2000). 6.

[2] For samples of recipes see *The Great Curries of India* by Camellia Panjabi, 40.

[3] For a more complete explanation of Dr. Rovesti's expedition, refer to Julia Lawless. *The Illustrated Encyclopedia of Essential Oils*, 18.

[4] See *The Columbia Electronic Encyclopedia*, Sixth Edition (Columbia University Press, 2003), s.v. 'Babur.'

[5] Sarina Singh. *India*. 10th ed. (Oakland, CA: Lonely Planet, 2003). 21-26.

[6] Ibid. 53.

[7] Roy E. Shepard. "History of the Rose." *Science Education* 39 (1954). 182-183.

[8] Attilio Petruccioli. "Shalimar Gardens in Srinagar."

http://archnet.org/library/sites/one-site.tcl?site_id=8861 Accessed 19 March 2007.

[9] Suni Systems (P) Ltd. "Chandigarh." http://www.webindia123.com/city/chandigarh/park.htm Accessed 19 March 2007.

[10] Eyewitness account from Raj Kumari Sharma, the author's mother. 14 February 2007.

[11] Lonely Planet Publications. "Lalbagh Botanical Gardens." http://www.lonelyplanet.com/worldguide/destinations/asia/india/bangalore?poi=1000151254 Accessed 14 April 2007. See also Singh. *India*. 53.

[12] This river is sacred to Hindus and, according to popular belief, to bathe in the sacred waters of the river is to wash away an individual's sins. Another belief states that scattering the ashes of the dead into the river will ensure that the soul reaches heaven. Several temples have been built at strategic points along its course in order to attract large numbers of pilgrims year-round.

[13] Eyewitness account by the author on a visit to India. 28 March 2004.

[14] Hindu philosophy describes four paths to God realization. These paths are Bhakti yoga, Karma yoga, Raja yoga and Hatha yoga. Bhakti yoga or marg (path) is the path of love or devotion to God. Karma yoga is the path of service. Raja yoga or royal (path) is the path of scientific meditation and Hatha yoga is the path of disciplining the body to transcend the body's consciousness.

[15] The Hindu trinity consists of Brahma, the creator; Vishnu, the preserver; and Shiva, the destroyer.

[16] Rosefarm.com. "A Brief History of Roses." http://www.rosefarm.com/history.php Accessed 15 March 2007.

[17] Marc Halpern. *Principles of Ayurvedic Medicine, Student's Textbook*. Part One. 6th ed. (Grass Valley, CA: California College of Ayurveda, 2004). 1-3.

[18] Frawley. *Ayurvedic Healing: A Comprehensive Guide*. 6.

[19] Vasant Lad. *Ayurveda, The Science of Self Healing*. (Delhi: Motilal Banarsidass Publishers, 1984). 18.

[20] Frawley. Ibid.

[21] Halpern. Ibid. 10-17.

[22] Ibid. 10-12.

[23] Frawley. Ibid. 327.

[24] David Crow. "The Pharmacy of Flowers." http://www.floracopeia.com/article.php?article=10 Accessed 17 March 2007.

[25] Eyewitness account by the author, 18 January 2007.

[26] Richard Burton, "On the Arts and Sciences to be Studied." *Kamasutra* [book on-line]. http://www.kamasutra-sex.org/text/kama103.htm Accessed 17 March 2007.

[27] Richard Burton. "On How to Begin and How to End Congress." *Kamasutra* [book on-line] http://www.kamasutra-sex.org/text/kama210.htm Accessed 17 March 2007.

[28] Crow, Ibid.

[29] Ibid.

[30] See *The Columbia Electronic Encyclopedia*. Sixth Edition. (NY: Columbia University Press, 2003). s.v. 'Attar of Roses.'

[31] Christopher McMahon. "More about Sacred Oils and Pure Essential Attar Oils © Exclusively at Erzulie's." http://www.erzulies.co.uk/site/articles/view/22 Accessed 17 March 2007.

[32] Thiru, Dayanidhi Maran. "Speech of Thiru. Dayanidhi Maran Hon'ble Union Minister of Communications and Information Technology at The Release of Stamps on 'FRAGRANCE OF ROSES'." Hotel Taj Lands, Mumbai, 2 February 2007. http://www.dmaran.nic.in/speechdisplay.php?id=207 Accessed 18 February 2007.

[33] Hindilyrix.com. "Gulaabi Aankhen." http://www.hindilyrix.com/songs/get_song_Gulaabi%20Aankhen.html Accessed 14 March 2007.

[34] See *Britannica Concise Encyclopedia*. (NP: Encyclopedia Britannica, Inc. 2006). s.v. 'mantras.'

5

Rose to Rosary: The Origins of the Flower of Venus in Catholicism

By Lisa Cucciniello

Centuries of Roman Catholics have shown their piety and dedication to the Virgin Mary by praying the rosary. Pope Leo XIII, who reigned from 1878-1903, referred to the rosary as "this pious method of prayer." Several decades later, one of Leo XIII's successors, Pope Pius XII, who reigned from 1939-1958, explained, "Such piety towards the blessed Virgin is the hallmark of a profoundly Catholic heart."[1] Throughout Catholic history, endorsement of the rosary by church leaders helped make this devotion an enduring Catholic icon. But the *origin* of the Catholic rosary and its connection to the actual flower, the rose, remain obscure. In fact, the rosary evolved over centuries, beginning in 753 BC with the founding of Rome and continuing to the present time when the late Pope John Paul II added his touch to this important symbol in 2002, causing some contention among Catholic congregations.

What began as a simple rose serving as a symbol of the pagan Roman goddess Venus ultimately transformed into a Catholic icon glorifying the Virgin Mary. Part of this metamorphosis included using both the rosary and the Virgin Mary as a theme in Medieval and Renaissance artwork. Publicizing the rosary would help promote the Catholic faith and maintain the dedication of the congregation, especially when the Catholic Church was in danger of losing members, such as during the Albigensian Crusade, the ravaging of Europe by the Black Plague and the Protestant Reformation. Active sponsorship of the rosary can be credited to the Dominican order of Catholic priests, an order that historically traces its roots to St. Dominic's preaching in the early twelfth century in the Languedoc area of Southern France.[2] The Dominican endorsement of the rosary is described throughout this chapter, with prominent Dominican leaders of the Catholic Church actively involved in the sponsorship of the rosary. In the fifteenth century, Dominican preacher Alan de la Roche vigorously promoted the rosary throughout France and the

Netherlands, deeming it his "special mission."[3] Contemporary to Roche was Father Jacob Sprenger, a Dominican priest who founded the first Confraternity of the Holy Rosary in 1475 in order to actively sanction commitment to the rosary, and Catholicism overall.[4] Along with a chronology of the Dominican approval of the rosary, this chapter also offers a possible explanation as to why the rose was chosen as an emblem to honor the Virgin Mary, even though this symbol is not unique to Christianity. Surprisingly, the rose as an icon of veneration has roots in the pre-Christian Era and was originally used by the Ancient Romans as a symbol of devotion to the Ancient Roman goddess Venus.

The founding of the City of Rome in 753 BC marked the beginning of the Roman Empire which would become one of the great civilizations of history. Until Emperor Constantine the Great Christianized Rome in the fourth century AD, Rome was pagan; its citizens worshipped more than one god, making them polytheistic.[5] An important symbol of Roman pagan rituals was the rose. The rose was the flower of Venus, an ancient Roman goddess derived from the earlier Greek goddess Aphrodite. Initially, Venus was the goddess of cultivated fields and gardens but would later be renowned as the goddess of love and beauty and, furthermore, revered for her wisdom.[6] As a result of Constantine Christianizing Rome, the Catholic Church became more prominent throughout Europe, and the rose would come to honor the Virgin Mary and her role in the life of Christ, thereby veiling its pagan roots.

It is at this time that Europe begins to see the adjustment of former pagan beliefs to fit a new Christian identity. The earlier pagan Roman rose would now be associated with the Virgin Mary as a result of the Christianization of Rome. However, according to John S. Stokes Jr., a Roman Catholic member of the Society of Mary, "The association of plants and flowers with the Blessed Virgin Mary originated with the early Church Fathers, who saw her prefigured in passages from the Old Testament containing nature imagery."[7] In the *Bible*, Canticles 2:1 states, "I am a rose of Sharon, a lily of the valley." Stokes further explains that "from this period, also, comes the legend that after Mary's Assumption into heaven, roses and lilies were found in her tomb."[8] As recently as 1955, Pope Pius XII, addressing a group of rose growers in Rome, discussed the pagan origins of the rose and its transformation into a symbol of Christianity stating, "When the memories of paganism were erased, the charm of the rose reverted to the true God."[9] The rose as a symbol of reverence underwent an evolutionary process that would transform it into the rosary. As the Catholic Church's power was threatened by heresy, disease, and, later, the Protestant Reformation, the rosary would be more aggressively promoted starting with the Albigensian Crusade in 1198.

For many years Catholic tradition traced the origin of the rosary to St. Dominic, founder of the Order of the Preachers or Dominicans devoted to ministering to and saving souls. Tradition acknowledged that in 1198, St. Dominic defeated the Albigensians, an heretical group that prevailed in southern France during the twelfth century whose teachings were contrary to those of the Catholic Church. According to the Albigensians, life was a dichotomy of good and evil, where good created the spiritual world and evil created the material world; thus, man was a living contradiction since he lived in the material world while seeking salvation in the spiritual one. Because of this conundrum, the Albigensians concluded that liberation of the soul was the true end of being, and the most commendable death was suicide by starvation. In contrast, the Catholic Church taught, and still teaches, that only God has "direct dominion over life"; therefore it is a sin to end one's time on Earth prematurely, such as by committing suicide.[10] According to Pope Innocent III who launched the Crusade against the Albigensians, the Albigensian heretics were more reprehensible than the Saracens, the Muslims whom the Catholic Church battled throughout the majority of the Crusades.

Though the Crusades were fought primarily against the Muslims in order to recapture Jerusalem, the Catholic Church waged war against the Christian Albigensians as well. The Catholic Church saw the Albigensian heresy as a force that would destroy the entire human race; hence, the group had to be abolished and a Crusade was launched to put an end to its heretical teachings. Catholic tradition taught that during the Albigensian Crusade, the Virgin Mary interceded on behalf of the Catholics as a result of St. Dominic praying the rosary, causing the Catholics to emerge victorious over the Albigensians.[11] Leaders of the church continually sanctioned this legend throughout history:

> The well-known origin of the Rosary, illustrated in the celebrated monuments of which we have made frequent mention, bears witness to its remarkable efficacy. For in the days when

the Albigensian sect, posing as the champion of pure faith and morals, but in reality introducing the worst kind of anarchy and corruption, brought many a nation to its utter ruin, the Church fought against it and the other infamous factions associated with it, not with troops and arms but chiefly with the power of the most Holy Rosary, the devotion which the Mother of God taught to our Father Dominic, in order [that] he might propagate by it. (Pope Leo XIII 1892)

However, the tradition of St. Dominic's connection to the rosary did not emerge until the latter fifteenth century. The legend began with Alan de la Roche, a distinguished member of the Dominican Order who was awarded the high honor of Master of Sacred Theology in 1473, a professional degree granted by Dominican seminaries.

Roche circulated the legend of St. Dominic and the conquest of the Albigensians to advocate the reputation of the Dominican order over the Carthusian order, another Catholic priestly community that committed much of its time to private prayer. In contrast, the Dominicans believed in active preaching to the outside community. Guy C. Bauman, writing in the *Metropolitan Museum Journal*, argues that an already existing strain between the two orders incited Roche to make a claim that would distinguish the Dominicans from the Carthusians.[12] Today, the tradition of St. Dominic and his defeat of the Albigensians is accepted as folklore, and some scholars, such as W.A. Hinnensbusch of The *New Catholic Encyclopedia*, now agree that the rosary's connection to St. Dominic is "apocryphal."[13] Perhaps the folklore was fostered because the Carthusian order had already begun to influence the Catholic faith by making daily devotion outside the formal mass more accessible to the illiterate population. It could easily be understood then why Roche would connect St. Dominic, the founder of the Dominican order of priests, to such an important event in Catholic history as the Albigensian Crusade.

Anne Winston, Associate Professor of German at Southern Illinois University at Carbondale, explains that Dominic of Prussia, a Carthusian monk from Poland noted for his intelligence and religious fervor in the early fifteenth century, predates Roche's claim that the origins of the rosary rest with St. Dominic. In his own work, *Liber*

Experimentiarum, Dominic of Prussia tells how he reflected on a series of meditations on the life of Jesus and the Virgin Mary while praying a succession of 'Hail Marys.'[14] During the time of Dominic of Prussia, the 150 Psalms of the *Bible* were regularly recited outside the formal mass as a form of devotional prayer. However, most people of the time were illiterate and thus many were excluded from this practice, as memorizing 150 individual Psalms without the aid of a prayer book was exceedingly complex. Consequently, some Catholics began replacing the 150 Psalms with repetitive 'Hail Marys' or 'Our Fathers' since these prayers were known by most Catholics and easier to commit to memory. This alternate form of recitation became known as the 'Psalter of Mary.'[15] According to the *New Catholic Encyclopedia,* Dominic of Prussia popularized this method of affection for the Virgin Mary as early as 1409, substituting the fifty Psalms of the *Bible* that spoke specifically of Jesus and the Virgin Mary with fifty 'Hail Marys.' This practice of saying fifty consecutive 'Hail Marys' became known as a *rosarium,* the Latin term for rose garden, which was often used during the fifteenth century to describe a succession of repetitive prayers.[16] By the time of Dominic of Prussia and Alan de la Roche in the later fifteenth century, an association between the rose itself and the Virgin Mary had already been established by the early church leaders, but further circulation of the rosary would help veil the earlier pagan connotations of the rose. Part of this process would be the transformation of the object used to show dedication. Pagan Romans used an actual rose, but Catholics grew to use mostly rosary beads for their devotional. However, the namesake of the Catholic rosary, the rose, remains.

A main component of the rosary dedication is the string of beads held in one's hands while the repetitive 'Hail Marys,' 'Our Fathers,' and 'Glory Bes' that make up the standard prayers of the Catholic rosary are recited. At one time, the individual beads were in the shape of little roses, the flower for which the dedication is named.[17] Each single bead represents a specific prayer of the rosary, and the prayers are said in a particular order according to the prescribed tradition of the devotion. As worshippers progress through the prayers, they pass one bead through their fingers for each prayer said, in order to keep count of prayers already recited. Like the rose, the tradition of using objects to keep track of prayers is not uniquely

Catholic. Though rosary beads are often associated with Western Europe and Catholicism, prayer beads date back to the third century in areas of the East, where they were used in other faiths such as Buddhism and Islam.

During the third century, Paul of Thebes, who is recognized as the first Catholic hermit, used stones to keep track of his daily ritual of saying three hundred 'Our Fathers.' According to Catholic tradition, Paul was born in Egypt and fled into the wilderness during the persecution of Christians by the Roman Emperors Decius and Valeranius circa 250 AD. He resided in a cave for ninety-one years where he devoutly practiced his faith until he died at the age of 114.[18] Saint Anthony, a wealthy man of the third century, used pebbles and knotted strings to keep track of his daily recitations when he was in the deserts of Egypt and Syria. When St. Anthony heard the words from the Gospel of Matthew, "If thou wilt be perfect, go and sell all thou hast," he felt they were directed to him and he renounced his worldly belongings to lead an austere life in the desert.[19] Marco Polo, a native of Venice who traveled to present day China in the thirteenth century, wrote of the Buddhist King Malabar wearing a chain of 104 gems which the king used to keep track of his daily devotions. Islamic tradition used strings of ninety-nine prayer beads to represent the ninety-nine names of Allah, a tradition that continues in the present day. This Muslim ritual appears to have its roots in the ancient Hindu god Shiva, who is worshipped as both the creator and the destroyer.[20]

Similar to these earlier traditions, Catholics used prayer beads to help keep track of recited prayers. However, Catholics eventually chose the term rosary beads instead of prayer beads, a reflection of their continual embrace of the primeval imagery of the rose. In addition to the name of the devotion that distinguishes this Catholic ritual from other traditions, the Catholic rosary consists of a series of fifteen meditations called mysteries, which are reflected upon as the devotee progresses through the prayers of the rosary.[21] These meditations focus on important events in the life of Jesus Christ and the Virgin Mary, and are taken from the Gospels of Matthew and Luke and the Acts of the Apostles. The reflection on these mysteries is fundamental to the Catholic rosary; many of the core beliefs of the Catholic faith, such as the death and resurrection of Jesus, are reflected upon throughout the recitation of the rosary. As people recite the standard consecutive prayers of the rosary, they are not just repeating prayer after prayer but also reflecting on the events of the mysteries and what their significance is to their faith.

As one prays the rosary and reflects on the meditations, he or she is taken through the crux of Catholic beliefs. Rev. Paul J. Oligny, who translated *The Holy Rosary: Selected and Arranged by the Benedictine Monks of Solesmes*, explains in the "Foreword," "Truly no more urgent and reliable education and re-education in the faith can be conceived than the rosary, the whole rosary that has become a habit. The Creed becomes what it truly is: a prayer. There we find the faith, whole and living, the true faith, the faith that begins with humility and ends in praise."[22] The Creed to which Oligny refers is the Apostle's Creed, a prayer believed to be composed by the original twelve apostles of Jesus forty days after His[23] death. Catholic belief states that on the fortieth day after Jesus rose from the dead, the Holy Spirit descended upon the twelve apostles and inspired them to publicly preach His mission. The Holy Spirit revealed Himself as tongues of fire over the apostles' heads, and at this time they composed the Creed. In doing so, each apostle contributed one of the articles, or beliefs, that are central to Catholicism.

To begin the prayer of the rosary, the devotee takes the rosary beads in his or her right hand and makes the sign of the cross. After the sign of the cross is made, the Apostle's Creed is recited, followed by one 'Our Father,' three 'Hail Marys' and one 'Glory Be.' The first mystery is read, followed by an 'Our Father,' a decade, or progression of ten consecutive 'Hail Marys' and another 'Glory Be.' As worshippers proceed through each decade, they reflect on the mystery and its significance assigned to that particular decade. This pattern of prayers continues for each of the mysteries. After the recitation of an entire set of mysteries, the rosary ends with the 'Hail Holy Queen,' an honorary prayer praising the Virgin Mary as the Mother of God and asking for her mercy on sinners. While there are fifteen mysteries in total, it is not obligatory to reflect upon all fifteen daily.

The *Catholic Encyclopedia* explains that certain sets of mysteries are reflected upon on certain days of the week. On Mondays and Saturdays one reflects upon the Joyful Mysteries. This series consists of the Annunciation, where the Virgin Mary was told by the angel Gabriel that she was going to bear Jesus; the Visitation, where the Virgin Mary visits her cousin Elizabeth, who in her old age, is miraculously bearing Jesus' cousin later known as John the Baptist; the Nativity, which is the story of the birth of Jesus, the Presentation of Jesus at the Temple, where in accordance with Jewish law, Jesus was presented at the Temple after being born; and the discovery of Jesus at the Temple, where Jesus is found at the age of twelve by His parents, preaching beyond His years to the elders of the Temple.

In 2002, an unexpected change was made to the centuries-old tradition of the rosary when Pope John Paul II added the Luminous Mysteries, a series of events taken from the four Gospels of Matthew, Mark, Luke and John that mark significant events in Jesus' public ministry. If you were to order the mysteries chronologically, according to when the events occurred in the life of Jesus, the Joyful Mysteries, which include Jesus' birth and childhood, would be followed by the recently added Luminous Mysteries, which detail Jesus' public ministry.

Marking his twenty-fourth anniversary as Pope, Supreme Pontiff John Paul II declared the year from October 2002 to October 2003 as "The Year of the Rosary" and requested the addition of five more events in the life of Christ to be added to the centuries old Catholic tradition of the rosary so that the prayer "could broaden to include the mysteries of Christ's public ministry between his Baptism and his Passion." Pope John Paul II asserted that it was important to include these events because "it is during the years of his public ministry that the mystery of Christ is most evidently a mystery of light." The pope wanted followers to reflect upon "certain particularly significant moments in his public ministry" before reflecting on the suffering of Jesus before His death.[24] The events that Pope John Paul II chose as "particularly significant" demonstrate the divine nature of Jesus within His own lifetime, making apparent that Jesus was indeed a super-conscious human being before His death and Resurrection.

An alteration of the prayer which had remained for the most part unchanged since 1569, when Pope Pius V is credited with standardizing the method of praying the rosary, was a surprise to many.

The pope's bold move of altering the rosary in 2002 risked agitating the more conventional members of the Catholic Church. BBC reporter Peter Gould explains "that tinkering with the wording would have risked offending traditionalists." Many Catholics were contented with their daily devotion as it was and were therefore somewhat resistant to the change of a prayer that had remained constant for generations. However, Pope John Paul II was well aware of the disparity this change could have caused and therefore left the decision of whether or not followers chose to reflect upon the newly added Luminous Mysteries to each congregation.[25] In a time of uncertainty among the congregation, the versatility of the rosary allowed this prayer to continue as a prominent Catholic icon.

This new series of meditations includes reflection on the Baptism of Jesus by His cousin John the Baptist in the River Jordan which, according to Catholic faith, marks the beginning of Jesus' public ministry; the Wedding Feast of Cana, where Jesus is said to have performed His first miracle by turning water into wine; The Proclamation of the Kingdom of God, where Jesus preaches of the Kingdom of Heaven; The Transfiguration, where Jesus took the apostles to a mountain top to pray and physically transformed in their presence; and the Last Supper, where Jesus gives the Holy Eucharist for the first time.[26] In the next set of reflections, the Sorrowful Mysteries, one would meditate upon Christ's Passion or the proceedings that led up to the Crucifixion of Jesus.

Believers contemplate the events of the Sorrowful Mysteries on Tuesdays and Fridays. This series of meditations includes the Agony in the Garden, where Jesus prayed in the Garden of Gethsemane shortly before His death and was betrayed by Judas; the Scourging at the Pillar, where Jesus is publicly beaten before a crowd; the Crowning with Thorns, where Jesus is mockingly crowned as the King of the Jews, yet not with a crown of gold and jewels as would be more common for a king at the time, but with a crown of thorns; the Carrying of the Cross, where Jesus carries His cross to the hill of Calvary, sometimes referred to as Golgotha where He was crucified; and the Crucifixion, where Jesus is nailed to the cross and dies. The Glorious Mysteries are reflected upon on Wednesdays and Sundays and include the Resurrection, where Jesus rose three days after He died; the Ascension of Jesus, where Jesus physically rose to heaven, body and soul; the Descent of the Holy Spirit, where the Holy Spirit appears to the Apostles, inspiring them to publicly preach His mission; the Assumption of Mary, where the Virgin Mary is taken body and soul to heaven, an occurrence that according to Catholic tradition, is unlike the experience of any other human being that has ever existed; and the Coronation, where the Virgin Mary is crowned queen of Heaven, again marking the Virgin Mary's special status in the eyes of the Catholic Church.[27] This pattern of mysteries most commonly recited by Catholics is called the Dominican rosary, bearing the name derived from the tradition of the rosary originating with St. Dominic. However, the Cistercians, an order founded by St. Robert, the Abbot of Molesme who felt that other orders of priests were becoming too relaxed with their rituals, reflected on a different series of events when they prayed the rosary.

The difference between the Cistercian and the Dominican rosary helps further demonstrate the metamorphosis of the Catholic rosary, including the shift in the overall focus of the prayer from a Marian devotion, a dedication to the Virgin Mary, to a series of reflections centered on the life of Christ.[28]

The Cistercian rosary places a greater emphasis on the life of the Virgin Mary than the Dominican tradition, setting the Annunciation of Mary as the eighth mystery - the first seven being centered on the life of the Virgin Mary as well. In contrast, the Dominican rosary sets the Annunciation of Mary as the first meditation, bypassing the other seven events in the life of the Virgin Mary, thus transforming the rosary from a prayer focused on the Virgin Mary to a Christ-centered reflection. The shift in the overall focus of the rosary could perhaps be explained by the fact that many of the core beliefs of Catholicism revolve around the Eucharist, the belief that the bread and wine offered at the Catholic mass become the actual Body and Blood of Jesus. This belief is reflected in the words said as a recipient receives the Eucharist at a Catholic Mass. As a Eucharistic minister[29] gives the Eucharist to the recipient, he says to the recipient, "Body of Christ." St. Thomas Aquinas, author of the *Summa Theologica*, one of the most complete works explaining Catholic doctrine, explains, "Since Christ's true body is in this sacrament ... it must be said then that it begins to be there by conversion of the substance of bread into itself."[30] Yet it should be noted that though the Catholic faith is centered on the Eucharist, that mystery was not added until 2002, the year in which Pope John Paul II added the Luminous Mysteries. In contrast to the Cistercian tradition, the Dominican rosary chronicles the life of Christ from the Virgin Mary's womb to His resurrection.

In the Dominican rosary, any events involving the Virgin Mary relate to her role in the life of Christ, making the meditations more of an "epic story" where each mystery leads into the next.[31] The story-like nature of the Dominican mysteries made the meditations easier to remember than the Cistercian rosary. The compact, portable nature of the beads and the simplicity of the prayers and meditations enabled people to recite the rosary anywhere or anytime, whether at home, working in the field, or on a long journey.[32] The ease of remembering the mysteries of the Dominican rosary, coupled with the compactness and portability of the string of beads, would help the rosary become a lasting icon of Catholic identity. But one must ponder why Roche, who lived in the mid-fifteenth century, would trace the rosary to St. Dominic and the defeat of the Albigensians three centuries earlier. Aside from trying to foster the Dominican reputation, part of the answer may be found in the events that surrounded the establishment of the Papal Inquisition, an organization established in the mid-thirteenth century, which the *Catholic Encyclopedia* defines as "a special ecclesiastical institution for combating or suppressing heresy."[33]

The century following St. Dominic's defeat of the Albigensians saw the establishment of the Papal Inquisition. The threat that the Albigensians posed to the Catholic Church during the twelfth century, along with the spread of Islam, impressed upon Church leaders the need for an institution like the Inquisition to avoid losing membership although "the positive suppression of heresy by ecclesiastical and civil authority in Christian society" predated this menace and is, indeed, "as old as the Church [itself]."[34]

By the time the Inquisition was established, the Crusades, a series of Holy Wars waged by the Catholic Church with the purpose of taking back the Holy Land from the Muslims, had been underway for over 150 years. Though the Crusades were mostly fought against Muslims, at times the Catholic Church would also combat Christians, as in the case of the Albigensians. The establishment of a formal Inquisition justified the "suppression of heresies," either Christian or Muslim, thereby legitimizing the Church's actions.[35] By using St. Dominic and his defeat of the Albigensians as an example, leaders of the Catholic Church such as Roche helped prove that Catholicism was the true faith.

By continually propagating the power of the rosary for the next several hundred years, the papacy was able to maintain dedication to the Catholic Church. Pope Leo XIII declared, "The Rosary was instituted chiefly to implore the protection of the Mother of God against the enemies of the Catholic Church, and, as everyone knows, it has often been most effectual in delivering the Church from calamities." Pope Benedict XV, Supreme Pontiff from 1914-1922, avowed, "In the struggle against the Albigensian heretics who uttered horrible blasphemies as they contested all the truths of faith … Dominic defended the sacredness of these dogmas with all his strength, and implored the help of the Virgin Mother by addressing this invocation to her very frequently: 'Grant that I may praise you, O holy Virgin; give me strength against your enemies'."[36] By using St. Dominic's victory as an example, the Catholic Church made it clear that anyone who spoke against its teachings would be defeated; thus anyone who was in contradiction of Catholic doctrine could indeed suffer the wrath of the Inquisition. As Roche began his proliferation of the rosary in the fifteenth century with the legend of St. Dominic, Father Jacob Sprenger, another member of the Dominican order, would join the fight against heretics as well.

In 1475, Father Sprenger founded the Rosary Confraternity, an organization that required one to register his or her name in a book and promise to recite the rosary on a daily basis. These registers were kept yearly, perhaps being used as a means to keep track of the Catholics in the areas which Confraternities were present. Later on, Jacob Sprenger would also write the preface and co-author *The Malleus Maleficarum*, which translates to *The Witches' Hammer*. For almost 300 years, this book would serve as a manual for the Inquisition on how to locate, torture, and kill a witch, with chapters such as "The Method of Passing Sentence upon one who hath Confessed to Heresy but is not Penitent." At the time of Roche and Sprenger, allegations of witchcraft were rampant throughout parts of Europe, especially targeting women, with tens of thousands being tried, convicted and executed as witches for practices such as using certain herbs to reduce the pain of childbirth. On 5 December 1484, Pope Innocent VIII issued a special bull, or church document, against witchcraft. This bull coupled with Sprenger's book would lead to mass hysteria throughout Europe, ending with the death of many women, some estimates as high as one million, who were allegedly guilty of witchcraft.[37] Perhaps if a person's name was not on the list of the Confraternity, that individual could be a potential target of the Inquisition and placed under suspicion of heresy or witchcraft, an accusation that seemed to fall upon women more than men as a result of Sprenger's book.

The model of St. Dominic and the rosary perhaps gave women a sort of protection from the witchcraft allegations that were so prevalent at the time. By practicing Catholic devotions such as the rosary, women might have been able to avoid being accused of sorcery. The example of the Virgin Mary gave women a model of piety to strive for. Later, in 1892, Pope Gregory XIII would remind followers, "In Mary, we see how a truly good and provident God has established for us a most suitable example for every virtue . ... We strive with greater confidence to imitate her."[38] Propagation of the rosary through the development of confraternities helped attract many followers, since being a member of a Confraternity had many benefits.

Joining a Confraternity was simple; protocol required one to sign his or her name in a register and pledge to recite the rosary. Being a member of a large praying community appealed to so many because members had the ability to offer prayers on behalf of the deceased as well as themselves.[39] In doing so, one could reduce his or her time or the time of a departed family member in purgatory, which according to Catholic belief, is a condition of temporal punishment; those who have died with unresolved sins remain in an intermediary state for a period of time prior to entering heaven. Prayers on behalf of those in purgatory allegedly reduced one's time there, a practice commonly termed 'an indulgence.'[40]

The relief from purgatory in addition to the almost open membership policy of rosary confraternities appealed to many, and within the first seven years of the establishment of Sprenger's Confraternity in Cologne in 1475, membership expanded to over 100,000, and Confraternities began appearing in other parts of Western Europe as well.[41] Such rapid growth can be attributed to the fact that Confraternities had few restrictions; there were no fees to join and everyone was welcome regardless of gender or social class. Other rituals of the Catholic Church, such as the Eucharist, excluded women. Priests were required to perform the blessings over the Bread and Wine of the Eucharist, and since women were forbidden to be priests in the Catholic Church, as they still are today, perhaps many women found consolation in the fact that they could fully participate in the rosary without the need for a priest acting as a liaison.

Little did many women know that the appealing nature that caused them to join the Confraternity to begin with might have been a way to keep watch over their activity within the Catholic Church. Joining the group that seemed to welcome them with open arms just might have saved their lives. Though both Sprenger and Roche collaborated to keep as many people faithful to the church as possible by advocating devotion to the rosary, they did not agree on using the word 'rosary' to represent this dedication, perhaps due to the pagan origin of the symbolism of the rose.

When Alan de la Roche began promoting the rosary in the fifteenth century, he was opposed to using the term 'rosary' because he felt it had "profane associations attached to the rose chaplet" and "worldly connotations." From Ancient Rome to the Middle Ages, a garden cosseted by a rose hedge had been an "ideal" place for romantic encounters. Suitors would give a rose circlet to the girls they wished to flatter, as it was said to denote reverence and respect.[42] Roche's resistance to a term that signified reverence and respect does not seem rational; could there be another reason why Roche was so steadfast about using Dominic of Prussia's term 'Psalter' instead of the term 'rosary'? Perhaps Roche was well aware of the ancient pagan correlation of the rose to Venus and was therefore wary about associating the rose with the Virgin Mary. The ancient Romans celebrated the 'rosalia,' a spring festival that honored the dead. However, even after Constantine the Great Christianized the Roman Empire, a variation of the festival continued into the Middle Ages; the festival was held in May when the roses began to bloom, and roses were worn as head-dresses at the celebration. This tradition would further be Christianized as well, giving Catholics the Feast of the May Crowning, a celebration that still continues in the present day. At the May Crowning, the Virgin Mary is honored by Catholic congregations who gather to pray the rosary as a community, and a statue in the likeness of the Virgin Mary is crowned with the circlet of roses.

Another example of the translation of pagan to Christian was the Ancient Greek's use of the red rose to symbolize the pierced foot of Aphrodite, the ancient Greek goddess who would later become synonymous with the Ancient Roman goddess Venus. Later, in the book of Wisdom of the Catholic *Bible*, this image would translate to 'The Precious Blood of Our Lord.' The Greek Aphrodite and the Roman

Venus, both ancient pagan idols, were revered for their wisdom as well as their love and beauty.

The connection of the rose to both Ancient Greek and Roman paganism may have been unsettling for Roche. At the time of Roche and Sprenger, the Renaissance, or rebirth of ancient learning and understanding, was underway. Some of the ideas experiencing revival did not comply with the teachings of the Catholic Church, as much of the knowledge was derived from Ancient Greek and Roman philosophies and culture, including pagan religious beliefs. Perhaps Roche feared that linking the Virgin Mary's dedication to the rose would too closely equate her with pagan rituals. Opposed to this correlation, Roche favored the use of the term "Psalter," the practice formally propagated by Dominic of Prussia. Sprenger, however, insisted on the term 'rosary' because the rose was a symbol already recognized by many.

Perhaps Sprenger rationalized that it might be easier to present people with an already familiar term and transform it into a Christian icon, as opposed to introducing a completely new term such as 'Psalter' to which the people could not easily relate.[43] Sprenger's approach indeed triumphed, giving Catholics a 'rosary' instead of a 'Psalter of Mary.' It seems a bit ironic that the same man who wrote the manual to seek out witches was sanctioning the connection of a pagan symbol to the Virgin Mary. It is almost as if Sprenger approved the remainder of a certain level of paganism, or at least enough to conveniently translate to Catholicism. Yet practices such as midwifery and treatment of ailments through the knowledge of herbs perhaps empowered women too much, and, therefore, such activities would cause many to be accused of being witches. If nothing else, Sprenger was a practical man and his plan achieved its desired goal; many remained faithful to Catholic teachings by praying the rosary. It seems fitting that the rose, a symbol formerly linked with Venus and Aphrodite, was chosen in Catholic tradition to represent the Virgin Mary. Venus, Aphrodite and the Virgin Mary have very similar characteristics, as each of the three patronesses was revered for her love, beauty and wisdom.

"The rose was known as the flower of Venus to the ancient Romans. Anything spoken *sub rosa*, or 'under the rose' was sacred and was not to be revealed to the uninitiated."[44] The term "uninitiated" refers to those outside the Cult of Venus, an ancient pagan

Roman following that was devoted to Venus and actively sanctioned by the Emperors of Rome beginning in the third century BC. The first temple in honor of Venus was instated on 18 August 293 BC. As Rome became Christianized in the third century AD, cathedrals were built in place of the temples dedicated to Venus, putting Mary in her 'Palaces of the Queen of Heaven,' where she was referred to as the 'Rose, Rose-bush, Rose-garland or Mystic Rose.'[45] The rose remains in the form of the rose window in gothic cathedrals such as Chartres and Notre Dame, both in France. This process of Christianization would endeavor to erase any residual images of the pagan Venus by replacing them with symbols of the Catholic Virgin Mary; however, many similarities between the two patronesses remained.

In Catholic tradition, the Assumption of Mary, celebrated on 15 August, is very close to the date of the inauguration of the first Temple of Venus. Catholic doctrine teaches that the Assumption of Mary is the Virgin Mary physically ascending to heaven, body and soul, to exist in the presence of God. This incidence is distinctive to the Virgin Mary, having never been experienced by any other human being.[46] At the time of her Assumption, the Virgin Mary's level of awareness was second only to that of her Son, and her graces gave her vision superior to that of any other "blessed" person. "She surpassed the other blessed in her knowledge of creatures, particularly in her knowledge of her fellow man."[47]

The Assumption of Mary, the fourth Glorious Mystery of the rosary, bears a striking resemblance to the astuteness for which the pagan Venus is revered. Both Venus and the Virgin Mary were venerated for their wisdom; it was a wisdom possessed by few, a wisdom not bestowed upon the "uninitiated." One might argue that the term rosary never truly adopted a new meaning when it became associated with the Virgin Mary as it maintained its original symbolism of wisdom in several respects. The word 'rosary' began to appear in German and Latin vernacular texts in about the thirteenth century. During this time, the German *rosenkranz* and the Latin *rosarium* often referred to an anthology of texts and were therefore appropriate terms to use to describe the Virgin Mary and her wisdom.[48] The wisdom of both the Virgin Mary and Venus could be seen as a kind of metaphorical anthology. So, although Sprenger advocated the term rosary based on its familiarity to the public, it can also be argued that the term maintained its original meaning as well. As

Venus was admired for her wisdom, so was the Virgin Mary at the time she was called to be with God. The similarities between the two women allowed for a suitable translation of the term rosary into the Catholic belief system, helping Christianize former pagan beliefs.

> Crimson rose of heavenly fragrance … Beautiful Mary, you raised women to a new dignity. Adam's Fall did not stain thee. Thy body did not know corruption. Peaceful sleep fell upon thee. Thy feet on the wings of angels. (Thomas Aquinas, *Sunday Sermon*)

As mentioned earlier, the connection between the rose and the Virgin Mary began with early prominent figures in the Catholic Church. St. Thomas Aquinas, born in 1225, wrote the *Summa Theologica*, a complete work on Catholicism that would be used both to instruct beginners and also to teach the proficient in the Catholic faith. This work would eventually earn him the title 'Doctor of the Church' and he often praised the Virgin Mary in sermons, at one time using the phrase, ave rosa spina careens, 'a rose without thorns.' Aquinas' reference to the Virgin Mary being 'without thorns' symbolizes the Catholic belief that the Virgin Mary was born without original sin, the sin that all others are born with as a result of Adam and Eve eating the fruit from the Forbidden Tree in the Garden of Eden. This seemingly miraculous state at the moment of the Virgin Mary's conception is celebrated yearly in the Catholic faith on 8 December, with the Feast of the Immaculate Conception. Once again the Virgin Mary is distinguished from other human beings, even from the moment she was conceived. Later on Peter Martyr, a member of the Dominican order in the thirteenth century who would eventually become the Inquisitor General of Italy, called the Virgin Mary *ave java mundi rosa*, or 'radiance of the world.' Similarly, St. Catherine of Siena, a Dominican nun who dedicated her life to helping the poor and the sick, praised the Virgin Mary as 'the rose of heaven.'[49] Roche and Sprenger publicized these already formed connections between the Virgin Mary and the rosary at the same time alleged witches threatened the power of the Catholic Church. At this time, the rosary was also a way to satisfy the needs of the people during the continued devastation of the Black Death, a plague transmitted by fleas and carried on the backs of rats that eliminated one third of Europe's population during the Middle Ages.

Though the Black Death made its first appearance in Europe in the fourteenth century, the plague periodically swept through the continent in waves, reappearing as late as the seventeenth century. At times, the plague was wiping out as much as one third of entire village populations.[50] Priests were able to offer little relief to the masses for several reasons; the first was that the Black Death knew no profession; many priests also fell victim to the disease, therefore leaving some towns without parish clergy. Another reason that priests could not help was because even if they were brave enough to face the sick, knowing the possibility that they could fall ill themselves, there was little assistance they could give without the aid of modern medicine. As a result, people began losing faith in the Catholic Church. The devastation of the plague caused people to desire a way to show their dedication to God without having to go to a priest, who served as an intermediary. At this point in European history, significant portions of society, such as the poor and the peasants who comprised the majority of the population, were either illiterate or not versed in Latin, the language of the Catholic Church at the time, and could not decipher written prayer books. The rosary provided a solution to all these issues while keeping congregations committed to the Catholic Church; it provided people with a more direct involvement in devotion to God with something they could easily remember and carry with them at all times, and also helped restore their faith in the Catholic Church.[51] Similarly, the practice of the rosary kept followers faithful to Catholicism during the Protestant Reformation, a movement that began in 1517.

Prior to the Protestant Reformation, some corrupt leaders of the Catholic Church were selling indulgences, allowing followers to believe they could reduce their time or a deceased family member's time in purgatory for a sum of money. Martin Luther, a Catholic monk, publicly opposed the Catholic Church's retailing of indulgences. As a result of Luther's protest, some Catholics completely broke away from Catholic teachings to follow those of Martin Luther and later became known as Lutherans. In response to the significant loss of its members, the Catholic Church promoted the rosary as an indulgence that would perhaps prevent followers from straying to the newly formed Lutheran sect of Christianity.

The Catholic Church was faced with a bit of a dilemma when dealing with the issue of purgatory and indulgences during the Protestant Reformation. The doctrine of purgatory was already cemented in the minds of the people; therefore, it would not have been feasible for the Church to renege on this teaching, but as a result of Luther's protest, followers were no longer left with a means to reduce their time. Since the sale of indulgences was causing the church to lose members, the Catholic Church needed to give followers an alternative means to limit residence in purgatory while preventing the further loss of members. The rosary, already a familiar practice to Catholics, could conveniently fill this need. As late as 1832, popes continued the proliferation of the rosary as an indulgence, including a statement by Pope Gregory XVI in his Apostolic Letter *Benedicentes*, which stated, "We have, therefore, decided to grant papal approval to this beneficial institution and to enrich it with indulgences." Throughout many parts of Europe, the rosary served as a way to keep followers faithful to Catholicism, for now Catholics had a way to reduce their time in purgatory without having to pay corrupt church officials. Moreover, in England the rosary would serve as one of the last remaining links to Catholicism for English Catholics.

In 1559, Queen Elizabeth I of England issued the Second Act of Uniformity, making Protestantism the solitary lawful religion in England.[52] By the 1570s, few Catholic priests remained in England and most Catholic images and vestiges in the country, including churches, were destroyed as a result of Anglicans, the followers of the Church of England, opposing what they considered the ritualistic nature of the Catholic mass. In spite of this hostility, devotion to the rosary actually increased. Because of its size and portability, the string of beads could be masked effortlessly in a man's pocket or sewn inside a woman's dress, therefore allowing Catholics to continue with devotion to their beliefs.

Following the Act of Uniformity, participation in the Eucharist was impossible in England because of the illegality of Catholic ceremonies. The rosary therefore became one of the only ways for English Catholics to express devotion to the faith.[53] According to Catholic tradition, lack of participation in the Eucharist would increase one's time in purgatory. As mentioned earlier, praying the rosary would help reduce a person's time there; as a result, the prayer of the rosary would help combat the trouble inflicted by the Protestants. A Catholic's time in purgatory was increased because of lack of participation in the Eucharist, but was reduced by praying the rosary and in fact, its popularity rose as a means of contesting Protestant actions against the Catholics of England. While the rosary is a devotion to the Virgin Mary, one accesses Christ through her, "the source of all saving grace."[54] On 1 September 1883, Pope Leo XIII explained in his *Supremi Apostolatus* the connection of the Virgin Mary to her son Jesus when she is prayed to for guidance. "She, who is associated with Him in the work of man's salvation, has favor and power with her Son greater than any other human or angelic creature has ever obtained or ever can obtain."[55]

The rosary allowed devotion to Catholicism to remain strong for Catholics in Elizabethan England. William Eric Brown, author of *John Ogilvie: An Account of his Life and Death with a Translation of Documents Relating thereunto*, explains the fate of John Ogilvie, an English Catholic who was executed for treason. Before his death, Ogilvie bid farewell to his fellow Catholics, taking his rosary with him to the gallows. As a final act, he threw it into the audience where it landed on an observer.[56] Though practicing Catholicism posed its risks, Ogilvie's bold action demonstrated the deep devotion that still existed for some remaining Catholics in Protestant England. Perhaps Ogilvie's last gesture was one of defiance. Since he was already destined for the gallows, possibly he took the risk knowing he had nothing to lose. Another explanation could be that he was, in fact, showing one final act of defiance by stating he was indeed still a faithful Catholic in a country that legally no longer allowed him to be. In England, Catholics attempted to cling to the rosary, as it was one of the last remnants of Catholicism in their Protestant country. However, in other areas of Europe, such as France and Spain where the Catholic Church still had the majority of followers, the rosary was actively endorsed.

As the Protestant Reformation continued to take members of the Catholic Church throughout parts of Europe, Pope Pius V and Pope Gregory XIII sanctioned the Feast of the Holy Rosary to be celebrated on 7 October 1571. On this date, the Muslims of the Ottoman Turkish Empire made one last attempt to expand their territory as far as Spain, but were defeated in the Battle of Lepanto. This victory over the Turks was credited solely to the Virgin Mary. It was said that as the battle was being fought, the Confraternity of the Rosary was meeting at the Dominican headquarters in Rome where they recited the rosary with the belief that doing so would lead to the defeat of Muslim enemies.[57] In the midst of the Protestant Reformation, the declaration of the Feast of the Holy Rosary reminded Catholics of their true faith. Perhaps the papacy used the defeat of the Turks as an example, helping show the importance of remaining true to Catholicism in the midst of others straying from traditional Catholic beliefs. With parts of Europe experiencing religious turmoil, both Catholics and Protestants were vying to be the dominant religion. The Feast of the Holy Rosary could have conceivably served as a reminder of the possible fate of non-Catholic believers, that they too would be defeated just as the Turks had been, and not just by an army such as that of Spain which defeated the Ottoman Turks, but by Divine Intervention, in much the same way as the Albigensians were allegedly defeated a few centuries earlier when being slaughtered during a Crusade. It is interesting to note that the defeat of the Ottoman Turks produced a result much like the defeat of the Albigensians of southern France. However, propagation from popes and other church leaders was not the only method used to foster devotion to the Virgin Mary's dedication. Artists, especially during the Medieval and Renaissance periods, played a significant role in the process as well.

Medieval and Renaissance art helped define a universal format for the rosary, allowing it to gain a foothold as well as helping sustain its general recognition. The invention of the printing press in 1452 made vernacular prayer books, books that were printed in the language commonly spoken in a particular area, more readily available. However, other alternatives were needed for the illiterate populace. To help the rosary gain popularity, artists began using it as a focal point for their work. Purgatory was a common theme; those unable to read popes' decrees could see the saving power of the prayers of the rosary beautifully depicted before them. Regardless of what language one spoke, anyone could appreciate their fate portrayed in artists' images.

In 1483, one of the most influential depictions of the mysteries of the rosary was created in woodcut for the brotherhood of Ulm by an artist named Dinkmut. No writing accompanied the work; therefore the message could reach all populations, including the poor, illiterate and those of various linguistic backgrounds. As late as the sixteenth century, this visual adaptation of the rosary was more popular than any written version. Pictorial versions of the rosary varied little from region to region, all portraying the mysteries of the Dominican rosary in some way. Each set of mysteries was often represented in a different color. The Joyful Mysteries were represented as a white rosary, detailing the birth and childhood of Jesus. The Sorrowful Mysteries were depicted in red, showing the passion of Christ. The golden rosary represented the Glorious Mysteries, signifying resurrection of Christ and the Assumption of Mary.[58]

In 1488, the Spanish artist Francisco Domenech created a woodcut called *The Fifteen Mysteries and the Virgin of the Rosary* which now resides in the Metropolitan Museum of Art in New York City. The painting consists of sixteen panels which portray the Joyful, Sorrowful and Glorious Mysteries of the rosary, as well as the Virgin Mary. To the right of the Virgin Mary is a kneeling nobleman who holds a string of rosary beads and three roses. This entire panel is surrounded by a string of rosary beads made from actual roses. The Virgin Mary is holding the Christ child, who is holding one of the flowers that makes up the perimeter of the panel. The nobleman has three assaulters in his wake, to whom he seems oblivious. Out of his mouth comes a stem of three roses and on his head rests a garland of roses.[59]

This panel of the Virgin Mary and the knight gives some insight into the early legends surrounding the rosary. In 1408, Vincent Ferrer, a Dominican preacher, attempted to unite the church leaders of Roman Catholicism and Eastern Orthodoxy, a sect of Christianity that observes the Byzantine Rite and follows the Patriarch of Constantinople as opposed to the papal rule in Rome. Ferrer told of a knight who began to recite the rosary after being captured. When his captors returned, they saw the Virgin Mary along with St. Catherine of Siena, a fourteenth century Dominican nun who had visions at an early age, and Agnes of Eulalia, a fourth century martyr who was murdered under the Roman Emperor Diocletian. Catherine held a plate of roses and Agnes held a needle and thread. As the knight prayed, each 'Hail Mary' produced a rose which the Virgin Mary strung into a crown and placed on the knight's head. The assaulters were astounded and inquired who the lady was; the knight was unaware that anyone else was present. The captors released the knight and decided to convert to Christianity.[60] For the illiterate population, a work that depicted imagery as powerful as this would give a prayer reference and a means of inspiration. The British Museum houses a work very similar to Domenech's which illustrates the mysteries as well.

The accompanying story to this version tells of a chap who would make a circlet of roses each day as a means of expressing his commitment to the Virgin Mary. Later he entered a Catholic monastery where he was unable to execute this daily custom. The leader of his order instructed him to say fifty 'Hail Marys' in place of the physical wreath he was accustomed to making. While traveling in the forest one day, he stopped to say his prayers but was interrupted by an assemblage of bandits that attempted to kill him. As they approached him, they saw a remarkable lady holding a chaplet; she was taking roses from the mouth of the monk and adding them to her wreath. She completed the wreath, placed it on her head and vanished. The captors had been watching close by and questioned the monk, who was unaware of the presence of the lady. They then set him free and converted, realizing the miracle they had witnessed.[61] Both versions of the legend are strikingly similar, although the paintings describing the stories are housed on different continents. Another illustration of the rosary, credited to Flemish artist Goswijn van der Weyden, also resides in the Metropolitan Museum of Art. This version is comparable to the already mentioned adaptations displayed at the Metropolitan Museum of Art and the British Museum.

In 1984, the Metropolitan Museum of Art received a series of sixteen paintings representing the fifteen mysteries of the rosary. The final panel shows the Virgin Mary in the center with St. Dominic on the left and a pope, an emperor and a king on the right, all facing a kneeling knight with three assailants behind him. The Virgin Mary is carrying the Christ child, who holds a rosary made of physical roses, fifty white roses representing the 'Hail Marys' and five red roses representing the 'Glory Bes' and 'Our Fathers.' This painting is thought to have been commissioned by a member of the Hapsburg kin, a royal German family whose members ruled throughout Europe from the late Middle Ages until the twentieth century.[62] The similarities between the works of Domenech, van der Weyden and the work in The British Museum give modern day devotees a glimpse of the fundamental aspects of the rosary. The fifteen mysteries are integral to the devotion and therefore appear in all three versions. The Virgin Mary is portrayed in each of them, which reinforces her significance to the Catholic faith. The fact that the rosaries are made of actual roses links the Virgin Mary to the wisdom for which she is revered. Although the above mentioned works of art show the Catholic link of the Virgin Mary to the rosary, a print, dated 1581, is housed in the New York Public Library and presents the Virgin Mary much in the same way Boticelli presents Venus in the fifteenth century work *The Birth of Venus.*

Roman mythology tells of the goddess Venus being born from the sea as a result of her angry mother Gaia, the goddess of Mother Earth, castrating her father Uranus, the god of the Sky and Heavens, and throwing his genitalia in the sea. As a result, her father's semen mixed with the foam of the sea and Venus was born as an adult emerging from a scallop shell and arrived on the shores of Cyprus. One should note that the Virgin Mary, as mentioned earlier, also has a special circumstance surrounding her birth. Though the Virgin Mary did not emerge from the sea, her state of freedom from original sin at the time of her birth sets her apart from any other human being. Venus was unique at the time of her birth because born as an adult from the sea. In Botticelli's painting, *The Birth of Venus*, the red-haired goddess appears to emerge from a giant shell and is surrounded by angel-like figures. Pink flowers, which appear to be roses, blow in the wind around her. An unaccredited print in the Spencer Collection of the New York Public Library titled *Prayers and Feast Days*, demonstrates the Virgin Mary emerging from nature in a similar manner. A red-headed Virgin Mary holds her son in her left hand, but appears to have no lower extremities; instead she exits from a giant rose. In her right hand, she holds a small rose toward which her naked infant reaches. A subtle glow illuminates her head. This sixteenth century work physically connects the Virgin Mary to the ancient symbol of wisdom, the rose. The *Catholic Encyclopedia* published in 1911 mentions that "Venus too often masquerades as the Madonna."[63] Madonna is another term often used to describe the Virgin Mary in the Catholic tradition.

The second half of Prayers and Feast Days consists of a woman with seven children, all joining hands to form a circle around her.

All faces are looking at the audience, and, in the background, a brick wall seems to enclose the group. The woman is flanked on both sides by roses.[64] Renowned for her knowledge, the Virgin Mary is surrounded by roses, the flowers synonymous with the wisdom of Venus. Revering the Virgin Mary's knowledge and its importance to the Catholic tradition, followers flocked to the Virgin Mary as children would to a teacher or a mother, as seen in *Prayers and Feast Days*.

Another unaccredited work in the New York Public Library, titled *Noble Milanaise*, dates from 1860-1861 and depicts a woman from the 1300s. While nobles in Medieval Europe were traditionally adorned in plush fabrics to display their affluence, this woman is veiled and layered in very simple fabrics. She wears a basic red tunic tied under her bosom; a light blue shawl, draped around her, reaches the floor. She has a white veil which follows her hair line, keeping her tresses covered but allowing her face to show. Her eyes are cast to the ground and gaze to the right, although it is not clear what she is looking at, as she is the only element present in this work. Her arms are folded and she is holding a rosary in her left hand. Her appearance is not one of wealth or distinguished bloodline, yet she is given the title "noble" by the unknown artist.[65] She is a rendition similar to many depictions of The Virgin Mary in Medieval and Renaissance art, such as Jean Bourdichon's "Miniature of Annunciation," another print housed at the New York Public Library.[66] The Catholic Church often preached that there was no better model of humanity than the Virgin Mary, the essence of spiritual nobility. By labeling someone dressed in the described manner as a noble, artists suggested to women to imitate this humble image. However, while visual art would serve an illiterate population in the Middle Ages and Renaissance Europe, more recent decrees from popes would help "foster" devotion among a more literate population.[67]

Reverend Paul J. Oligny translated a letter written by Pope Gregory XVI in 1832 that avowed, "The rosary is not only right because of the words, but also because it brings people closer together as they pray it [and] therefore God finds it favorable."[68] Pope Leo XIII, who would later become known as the 'Pope of the Rosary,' went on to issue nine encyclicals or official letters specifically referring to the rosary. On 1 September 1883, he issued his Encyclical *Supremi Apostolatus* which declared:

We consider that there can be no surer and more efficacious means to this end than by obtaining through devotion and piety in favor of the Virgin Mary, the Mother of God, the guardian of our peace and the minister to us of heavenly grace, who is placed on the highest summit of power and glory in heaven in order that she may bestow the help of her patronage on men who through so many labors and dangers are striving to reach the eternal city. (Pope Leo XIII, 1883)

Leo XIII goes on to remind his readers of the long-standing tradition of the rosary in the Catholic faith, labeling the heretics of both the Albigensian Crusade and Battle of Lepanto as the "dangers." As recently as 1955, Pope Pius XII addressed an international group of rose growers in Italy, commenting on the importance of the symbolism of the rose in Catholicism. He explained that it was the first Christians, not Roche alone, who rejected the rose because it represented "a life which they abhorred."[69] Nonetheless, the rose, once allegedly rejected by Christians for being too pagan, would later adorn some of the most famous Medieval Cathedrals, such as Notre Dame, in the form of the stained glass rose window. Pius XII continues to explain that the rose's connection to the Virgin Mary stems from St. Bernadette's apparitions. St. Bernadette of France was said to have a total of eighteen visions of the Virgin Mary at Lourdes, France. Many times, there were others around but Bernadette was the only one who could see the visions.[70] According to Pius XII, "It is fitting that the most beautiful of flowers is an offering to the most beautiful of creatures."[71] While Pius XII would actively remind the Catholic congregation of the significance of the rose throughout the history of their faith, Pius XII's successor, Pope John XXIII, would call an entire council that would alter the status of the rosary.

Earlier history shows the rosary being used at times to fulfil the needs of the illiterate population as well as to maintain the faith of Catholics when the church was in danger of losing followers. But during the reign of Pope John XXIII from 1958 to 1963, illiteracy was no longer a problem facing the Catholic Church; followers were, however, looking for a more active role in church rituals. The church officials of John XXIII's time sought to embrace the problems of the contemporary world. John XXIII called the Second Vatican Council in 1962, which helped church leaders address the spiritual needs of twentieth century Catholics.[72] Catholic doctrine was reexamined in order to better suit twentieth century congregations. Prior to this Council, Catholics were not actively encouraged to read the Bible, as it was believed that lay people could not fully understand the meaning of Scripture. The term 'lay people', or 'laity,' refers to those who are not of a particular religious order the way priests and nuns are but still identify themselves as adhering to the teachings of a particular religion. Prior to the Council, it was believed that the laity needed a priest to interpret the Gospels in a Homily, a sermon given by the priest at a Catholic Mass in which the priest relates the Scripture reading to everyday life. However, the Second Vatican Council realized it was beneficial to actively encourage people to read Scripture outside of the mass as well, as the general congregation was literate and much more educated than medieval followers. Though the Homily still remains a part of the Catholic Mass today, the need for the rosary as an alternate prayer form was no longer as urgent after the Second Vatican Council as it had been during the time of the Black Plague and the Protestant Reformation. People now began reading Scripture as a way to show their devotion to God as well. That is not to say that the Second Vatican Council discouraged dedication to the rosary; far from it. The Council simply offered the general congregation another alternative of devotion outside the formal Catholic Mass.

With one billion Catholics in the world today, devotion to the rosary remains, but the desire for an alternate prayer form is not as vital as it was during the Middle Ages.[73] Many remaining devotees were raised before the assembly of the Second Vatican Council, with the rosary being a vital component of their lives. For this generation of followers, the rosary was already an integral part of their daily ritual; therefore, they were comfortable with the practice and decided to continue with their devotion to the rosary. Catholics worldwide use the same strings of beads and reflect on the same mysteries when they are praying the rosary. However, there are some slight variations to the overall ritual. When the mysteries are read, which prayer ends the devotion depends on the country in which one prays. As in the time of Roche and Sprenger, the flexibility of the devotion to the rosary continues to appeal to many, helping it remain a lasting Catholic icon.

In the United States, the rosary ends with the 'Hail Holy Queen.' In Ireland, the rosary ends with the 'Litany of the Blessed Virgin.' In Uganda it ends with the 'Memorare.'[74] Though the ending of the devotion varies from one country to another, the central prayers of the rosary and content of the mysteries remain the same, as they are the core of the rosary dedication. Though there are some minor differences, the content of the mysteries remains unchanged. The mysteries, taken straight from the events of the *Bible*, are the crux of both the rosary and the beliefs of Catholicism, unwavering throughout centuries of the Catholic tradition.[75] When read in the United States, the leader of the prayer announces the mystery before each decade of 'Hail Marys' begins. In German-speaking countries, however, the mystery is said after the line in the 'Hail Mary,' "thy womb Jesus." For example, the fifth Sorrowful Mystery would read, "thy womb Jesus, who was crucified for us."[76] Just as the paintings created by different artists and housed in various museums maintained the central elements of the Catholic rosary in a consistent format, so, too, do devotees around the world who continue to recite the rosary, preserving it as a lasting symbol of Catholic faith.

For Catholics, the 1682-year-old tradition of the rosary, which began with Constantine the Great's Christianization of Rome, has incessantly served as a devotion to and a model of the piety of the Virgin Mary. Though the official origin of the rosary remains a topic of discussion, the significance of the rosary's connection to the rose continues to linger. While Alan de la Roche attributed the entire tradition of the rosary to St. Dominic, the significance of the rose actually pre-dates Christianity with its roots lying in the earlier pagan Roman goddess Venus. However, the formal recognition and spread of the rosary can be credited to the Dominican order, where members of the order such as St. Catherine of Siena, St. Peter Martyr and Alan de la Roche would actively preach the importance of dedication to the rosary. The most active endorsement occurred during times of both political and religious turmoil, such as the Inquisition, the Black Plague, and the Protestant Reformation, all periods in which the Catholic Church was in danger of losing members of its congregation. Thus the rosary has been used in the Catholic tradition in more ways than one, helping demonstrate the endless flexibility of the rose as a timeless symbol. Aiding in this diffusion of the rosary was Medieval and Renaissance art, which helped establish a universal format for rosary devotion. Most recently five more meditations were added to this centuries old tradition, with the late Pope John Paul II's Luminous Mysteries in 2002, showing the continual flexibility of the rosary right up to the present day. And yet, in all endorsements of this popular Catholic prayer, the uniqueness of the Virgin Mary is always remembered. Entire devotions to the Virgin Mary such as the rosary, prayers like 'Our Lady of Guadalupe' that end with the line, "Mystical Rose, pray for us," and works of art such as Domenech's "Fifteen Mysteries" and the "Virgin of the Rosary," continually embrace the symbolism of the rose and the ancient wisdom possessed by the Virgin Mary.

References

[1] Pope Paul VI, Pope John XXIII, Pope Leo XIII. *Seventeen Papal Documents on the Rosary.* (NY: Daughters of St. Paul, 1980) 111. *The Holy Rosary: Selected and Arranged by the Benedictine Monks of Solemes.* Trans. Rev. Paul J. Oligny, OFM (St. Paul Editions: Daughters of St. Paul, 1980), 230.

[2] See *Catholic Encyclopedia*, Volume XII, 1911, s.v. 'Dominicans.' The *Catholic Encyclopedia* is a compilation of over 10,000 articles which, as its preface states, "proposes to give its readers full and authoritative information on the entire cycle of Catholic interests, action and doctrine." It does not limit itself to Church doctrine, but also to significant contributions such as art and charity work made by Catholics throughout history. Dominicans are one of the many orders of priests and nuns in the Catholic Church. For a more comprehensive list of Catholic religious orders, see *A Guide to Religious Ministries for Catholic Men and Women*, an annual publication by the Catholic News Publishing Company, which enumerates with a brief overview of the history of the many orders of priests and nuns in the Catholic faith.

[3] See *Catholic Encyclopedia*, Volume I, 1907, s.v. 'Alan de la Roche.'

[4] See *Catholic Encyclopedia*, Volume III, 1912, s.v. 'Confraternity of the Holy Rosary.'

[5] The term 'Constantine the Great' is used by the *Catholic Encyclopedia*, Volume IV, published in 1908. In 325 AD, Constantine called the Council of Nicea, which helped unify Catholic doctrine, discussing issues such as the marital status of priests, excommunication, and the core beliefs of Catholicism. According to the *Catholic Encyclopedia*, Volume XI, published in 1911, paganism is any religion outside of Christianity, Judaism and Islam.

[6] *The Columbia Electronic Encyclopedia*, Sixth Edition. (NY: Columbia UP, 2003).

[7] John S. Stokes, "Flowers of the Virgin Mary," *AVE, Society of Mary.* (London, 1984) 1.

[8] Ibid.

[9] Pope Pius XII, "To an International Group of Rose Growers in Rome," *The Pope Speaks* 2 (Summer 1955) 133-135.

[10] See *Catholic Encyclopedia*, Volume XIV, 1912, s.v. 'Albigensians.'

[11] The *Catholic Encyclopedia*, Volume I, 1907, s.v. 'Crusades.'

[12] Guy C. Bauman, "A Rosary Picture with a View of the Park of the Ducal Palace in Brussels, Possibly by Goswijn van der Weyden," *Metropolitan Museum Journal*, 24 (1989), 138. The *Catholic Encyclopedia*, Volume III briefly explains the Carthusian order as being a small, slow growing order of monks whose daily life was greatly comprised of solitary meditation and reflection. In contrast, the Dominicans were an active preaching order dedicated to the salvation of souls.

[13] W.A. Hinnensbusch, "The Rosary," *The New Catholic Encyclopedia*, Boston. *Seventeen Papal Documents on the Rosary*, Daughters of St. Paul, 667.

[14] Anne Winston, "Tracing the Origins of the Rosary: German Vernacular Texts," *Speculum* Vol. 68 (Jul. 1993) 622.

[15] See *Catholic Encyclopedia*, s.v. 'rosary.'

[16] *New Catholic Encyclopedia* Volume XII (NY: McGraw Hill Book Company, 1967) 669.

[17] Today the rosary is most commonly made of oval or round beads, most likely for economic purposes. While the beads are a single string, terms such as "a pair of rosary beads," "a set of beads," or "a pair of beads" are sometimes used. The origins of the phrases are uncertain, but perhaps they refer to the fact that the rosary beads branch out from a common point, a crucifix, then split into two and connect to form a circle.

[18] Bauman, "A Rosary Picture with a View of the Park of the Ducal Palace in Brussels, Possibly by Goswijn van der Weyden."

[19] See *Catholic Encyclopedia*, s.v. St. Anthony.

[20] *Encyclopedia Britannica* 2006 explains that the word *Allah* is the standard Arabic word meaning God. Some of the ninety-nine names include, "the One and Only," "the Living One," and "the Real Truth." Anne Winston-Allen. *Stories of the Rose: The Making of the Rosary in the Middle Ages* (University Park: The University of Pennsylvania Press, 1997), 14.

[21] In 2002, Pope John Paul II sanctioned five more events in the life of Christ which he called the Luminous Mysteries. For the purpose of this essay, mainly the Joyful, Sorrowful, and Glorious Mysteries will be discussed.

[22] *The Holy Rosary: Selected and Arranged by the Benedictine Monks of Solemes*, trans. Rev. Paul J. Oligny, OFM, 14. Oligny states that an entire education of the faith can be learned from the devotion of the rosary because of the prayers, meditations, and beliefs expressed in the Creed. See *Catholic Encyclopedia* Volume I for a detailed explanation of the Apostle's Creed.

[23] In Catholic tradition, when referring to God or Jesus, using words such as Him, His, or He, the denotations are capitalized, as explained by the *Catholic Encyclopedia*, s.v. 'Jehovah,' because they refer to the sacred name of God or Jesus.

[24] For Pope John Paul II's complete address on the subject of the addition of the Luminous Mysteries, see *Apostolic Letter Rosarium Virginis of the Supreme Pontiff*, 16 October 2002, where Pope John Paul II addresses the Bishops, Clergy and Catholic Congregation on the subject of the Holy Rosary.

[25] For the complete article about Pope John Paul II's addition to the rosary, see *BBC News World Edition*, 16 October 2002, "The Pope's Belief in Daily Prayer."

[26] Catholics consider the Last Supper to be the First Holy Eucharist.

[27] For a complete description of the mysteries of the rosary, see *The Catholic Encyclopedia*.

[28] The *Catholic Encyclopedia* explains that St. Robert began the Cistercian order in order to return priests and religious to the teachings of St. Benedict. St. Benedict, a monk born in the latter fifth century, was renowned for his spiritual wisdom and is credited with being the founder of monasticism in the Catholic Church, a practice that requires one to live under strict religious vows.

[29] A Eucharistic minister is one who distributes the Eucharist to the congregation after it has been consecrated by the priest.

[30] *The Summa Theologica of St. Thomas Aquinas*, trans. Fathers of the English Dominican Province, (Second and Revised Edition: 1920). The *Catholic Encyclopedia* explains that the Eucharist is the actual presence of Jesus in the bread and wine. At a Catholic mass, the priest prays over the bread and wine, and through the process of transubstantiation, the bread and wine become transformed into the Body and Blood of Jesus.

[31] Anne Winston, "Tracing the Origins of the Rosary: German Vernacular Texts," 628-632.

[32] Ibid.

[33] The *Catholic Encyclopedia*, Volume VIII, 1910, s.v. "Inquisition."

[34] Ibid.

[35] See *Catholic Encyclopedia*, s.v. 'Inquisition,' for a complete description of the Inquisition and its role in the church.

[36] Pope Paul VI, Pope John XXIII, Pope Leo XIII. *Seventeen Papal Documents on the Rosary*, 103. Rev. Paul Oligny, trans. *Papal Teachings: The Holy Rosary* (Boston: The Daughters of St. Paul, 1980) 164.

[37] See *Catholic Encyclopedia*, s.v. Pope Innocent VIII.

[38] Pope Paul VI, Pope John XXIII, Pope Leo XIII. *Seventeen Papal Documents on the Rosary*. 118-119.

[39] Winston-Allen, *Stories of the Rose*. 111-120.

[40] The granting of indulgences continued in the Catholic Church throughout history. In 1832, Pope Gregory XVI issued his Apostolic Letter *Benedicentes*, which discussed how God found the rosary a pleasing dedication and was therefore allowing it as a form of indulgence.

[41] Winston-Allen, Ibid. 4.

[42] Historians continually debate the dates of the time period referred to as the Middle Ages or Medieval Europe. Some place it as early as 500 AD to as late as 1500 AD. However, the *Catholic Encyclopedia* begins the Middle Ages with the fall of the Roman Empire in 375 AD and ends it with Renaissance Italy in the mid-fifteenth century. Bauman, "A Rosary Picture with a View of the Park of the Ducal Palace in Brussels, Possibly by Goswijn van der Weyden," 150.

[43] Winston-Allen, Ibid. 101-108.

[44] Barbara G. Walker, *The Woman's Encyclopedia of Myths and Secrets*. (San Francisco: Harper Collins, 1983) 866-867.

[45] Ibid.

[46] Richard McBrien, *Catholicism: New Study Edition*, revised and updated. (San Francisco: Harper Collins, 1994), 1101-1102.

[47] *New Catholic Encyclopedia 2nd* Ed. (Washington DC: Catholic University of America, Thomson and Gale, 2003).

[48] Josef Klapper, "Mizellen: *Central German texts from Breslauer Handwriting*," Magazine for German Philology, 47 (1918), 83-87.

[49] References to St. Peter Martyr, St. Dominic and St. Catherine of Siena were found at NewAdvent.org, an online compilation of over 11,000 articles from *The Catholic Encyclopedia, Summa Theologica, Church Fathers, Bible,* and *How to Recite the Holy Rosary.*

[50] Black Death, *The Concise Oxford Dictionary of Archaeology.* (NY: Oxford UP, 2003).

[51] Anne Winston, "Tracing the Origins of the Rosary: German Vernacular Texts," 635.

[52] The first Act of Uniformity was issued by Henry VIII, the founder of the Church of England. After his death, there was a continual power struggle between Catholic and Protestant rulers, such as that of "Bloody Mary" the queen who wanted to take revenge on the Protestants, and therefore sanctioned their slaughter. As a result, devotion to Protestantism grew, defeating Mary's purpose of reinstating Catholicism as the faith of England. When Elizabeth came to the throne, religious turmoil was tearing her country apart; she therefore decided to unify England under one faith.

[53] Lisa McCain, "Using What is at Hand: English Catholic Reinterpretations of the Rosary, 1559-1642," *The Journal of Religious History* 27 no. 2 (2003): 161-176.

[54] Ibid. 162-165.

[55] Pope Paul VI, Pope John XXIII, Pope Leo XIII. *Seventeen Papal Documents on the Rosary.* 99.

[56] William Eric Brown, *John Ogilvie: An Account of his Life and Death with a Translation of Documents Relating thereunto* (London: Burns, Oates, and Washburn, Ltd. 1925) 144.

[57] Joseph Therese Agbasiere, "The Rosary: its history and relevance." *African Ecclesial Review* 30 no. 4 (2006) 250.

[58] Winston-Allen, *Stories of the Rose.* 8-32.

[59] Francesco Domenech, *The Fifteen Mysteries and the Virgin of the Rosary.* Engraving, 1488, The Metropolitan Museum of Art, New York.

[60] Valerio Serra and Boldu, *Libre d'or del rosary a Catalunya* (Barcelona, 1925) 22. See *Catholic Encyclopedia* for detailed explanation of St. Vincent Ferrer, St. Catherine of Siena and St. Agnes.

[61] Bauman, "A Rosary Picture with a View of the Park of the Ducal Palace in Brussels, Possibly by Goswijn van der Weyden," 24:141.

[62] Ibid. 135.

[63] The *Catholic Encyclopedia*, Volume XII, 1911, s.v. "Venus."

[64] *Prayers and Feast days*, 1581, NY Public Library, The Spencer Collection, New York.

[65] *Noble Milanese*, 1860-1861, NY Public Library, Mid-Manhattan Picture Collection, New York.

[66] Jean Bourdichon, *Miniature of Annunciation*, 1457?-1521?, NY Public Library, Medieval and Renaissance Illuminated Manuscripts from Western Europe, New York.

[67] *The Holy Rosary: Selected and Arranged by the Benedictine Monks of Solemes*, trans. Rev. Paul J. Oligny, OFM, 31.

[68] Ibid.

[69] Pope Pius XII, "To an International Group of Rose Growers in Rome," *The Pope Speaks* 2 (Summer 1955): 133-135.

[70] Today a Cathedral stands, said to be instructed by the Virgin Mary, and Lourdes,

France, remains a popular site for Catholic pilgrims.

[71] Pope Pius XII, "To an International Group of Rose Growers in Rome," 134.

[72] See *The Catholic Encyclopedia* for a complete history and description of the Second Vatican Council.

[73] McBrien, Catholicism. 1067-1071.

[74] See The *Catholic Encyclopedia*. The 'Litany of the Blessed Virgin' is one of the many prayers in honor of Mary. The 'Memorare' is often attributed to the French priest Claude Bernard but dates before his time. All three prayers are prayers said to honor the Blessed Virgin.

[75] Agbasiere, "The Rosary: Its History and Relevance." 30:250.

[76] *New Catholic Encyclopedia* Volume XII. (NY: McGraw Hill, 1967) 667.

6

THE ROSE AND ASTROLOGY

By Montgomery Taylor

Astrology is, first of all, a language of symbols. Although vulnerable to misinterpretation, those symbols communicate truth beyond the capability of words and, when carefully reflected upon, can help us understand who we truly are. The following essay is intended to explain, in simple terms, how the field of astrology or the study of the influence of heavenly bodies on human affairs is experiencing a renaissance and rebirth as a tool for psychological insight and for the forecasting of events and conditions. After suffering the condemnation of man's primitive understandings of the material universe, the great Swiss psychologist Carl Gustav Jung, 1875 – 1961, did much to restore the value of astrology as a means of more clearly delving into the individual and collective subconscious. Many of the world's most esteemed astrologers cite the work of Carl Jung, pointing out that symbolic realities are just as important to human nature as are physical realities, and maybe more important.[1] Ironically, Jung, as a master astrologer, often did astrological charts for his most problematic patients with great results. Astrology, according to Jung, was actually the oldest form of psychology because through astrology linear mental thinking can be aligned with intuition and emotional patterns.

An effective analogy that can help in understanding astrology is to see planets in the same manner as we view satellites. Planets are indeed satellites that orbit the sun along with the earth. When we view planets from the earth's perspective, we are actualizing what is called geocentric astrology, that is, planetary patterns in the heavens as viewed from earth. Just as our manufactured satellites have specialties such as weather, communications, military, etc. so do planets have specialties in terms of abstract cosmic energy that affect earthly events, including the human psyche. These energies are best represented in terms of human understanding as archetypes.

Jung put forth the concept that the gods and goddesses of ancient mythology throughout the world were symbolic in their nature and then shared by all humanity. Jung likewise postulated the existence of a level of consciousness shared by all, which he called the "collective unconscious."[2] Myths from every culture deal with the same subjects of mortality and immortality. The stories vary, but the messages are consistent, giving human beings a perspective on their relationship to universal consciousness. The whole point of astrology is to bring the unconscious information to conscious awareness.

Finally, true astrology, as opposed to the entertainment found in popular newspapers, is being given a place of respect in universities and schools of higher learning. The world's leading astrologers have founded professional research organizations such as the National Council of Geocosmic Research and others to assure the highest integrity and training of its practitioner members. Many consider astrology to be the earliest attempt to explore psychological and spiritual matters. After all, it was three astrologers that predicted the birth of Jesus! And to this day, the celebration of Easter is calculated as "the first Sunday after the first full moon, after the sun has entered the constellation of Aries." This explains why Easter is celebrated on a different date every year, but the other holidays remain the same.

Having a horoscope or natal chart calculated for the average person is a very recent development of the last century. Throughout history to the present day, royalty and world leaders from Queen Elizabeth I of England to presidents of the United States have taken astrological advice into consideration. In fact, since most of the lower classes had no idea of the exact time and date of their birth, horoscopes were usually cast only for royalty and heads of state or members of the aristocracy. In the secret writings of Freemasonry, it is well-known that Benjamin Franklin was a master astrologer and actually used the science of astrology in choosing the date for the signing of the Declaration of Independence.[3]

By the mid-20th century, a team of expert French researchers in statistics, Michel and Françoise Gauquelin, had spent years compiling data with a view of disproving astrology. To their great surprise, however, their research resulted in an unquestionable validation of this ancient science. Thousands of birth dates and related personal information fed into their computer convinced the Gauquelins that an uncanny consistency existed, that the movements of planetary bodies played a major role in identifying character, that those movements would lead a person to have distinct identifying traits and correlate eventually to certain career paths. As a result of their research, British astronomer Percy Seymour developed the theory of celestial magnetism to account for the influence of planets on human nature.[4]

Metaphysics and Quantum Physics

We may look at metaphysics as a poetic expression of quantum physics, where spirit and science validate each other in different languages. During these times of expanding awareness and curiosity, many polls reveal a growing interest in the mystical and unexplained.[5] If we approach spiritual experience with a perspective that is solely based on skepticism and doubt, nothing can be learned, and we doom ourselves to look at higher Universal awareness as nothing but a myth. This attitude became prevalent and imbedded during the seventeenth century, a period historically referred to as the "Age of Reason"; humankind wanted to see the world through the eyes of pure mental logic and linear reasoning. We quite understandably began the modern age with an overly simplistic and materialistic view of our universe.[6]

Granted, much of the pure wisdom and search for an understanding of humanity's relationship to the universe as found in the ancient mystery schools left a legacy that was distorted and devolved during the medieval period. It became corrupted, misinterpreted, and wrongly translated; it also fell victim to charlatans, religious zealots and condemnation by totalitarian religions masking as spirituality.[7]

Consider a rather simplified but effective explanation of how our rich and varied heritage of myth can be compared to a hypothetical but not far-fetched example of a modern-day airplane pilot encountering a Stone Age culture. Imagine your helicopter or small engine plane making an emergency landing in the darkest depths of the Amazon jungle. The local natives look at you with fear and wonder. Your appearance is different from theirs and you landed in their midst in a great bird or flying dragon from the heavens. They have never seen anything like you before. How will you communicate with them? How will you explain your technology? Then your cell phone rings, complete with talking pictures. You are telling the caller that you have landed safely, and you give your location so that you can be rescued. In the meantime, you try to explain your presence and your technology. But how can you do that? You have to invent a story that explains aerodynamics and reassures the natives that you do not have a god or a disincarnate slave trapped in your cell phone. They marvel at their new understanding that you give them in the form of your own allegorical myth, which tries valiantly to represent the facts that are natural and normal to you but are otherworldly to them. So it is with our astrological myths. The ancients of every culture could not help but notice that certain circumstances and events occurred with regularity whenever the sun, moon, or certain planets were in places that formed specific relationships to each other when observed from the earth.

And thus were born the "archetypes" – the characters that are defined in the dictionary as "a first form or model; the original pattern or model after which a thing is made."[8] The leading characters, or archetypes, of our astrological myths were merely attempts to put an abstract cosmic effect into story form to help understand and study the concept of planetary influence on our world. This way, the concept could be personalized and given life in our view of reality, and the evolved and unevolved expression of this energy or force could be attributed to the characters or archetypes in different stories that could help clarify the understanding of our relationship to both ourselves and the outer world.

In the most elementary terms, the natal birth chart is called a horoscope, Greek for "picture of the hour" for the time of birth of an individual and can represent the sky as seen from the date, time and location of a person's birth as well as the birth of a country, corporation, marriage, or any event for which information is needed. This "picture of the hour" is composed of planets, a Greek word for "messengers," which are energy vortex points that seem to have specialty functions as observed consistently for over 10,000 years. These planets are located in certain star constellations that comprise the "zodiac," Greek for "wheel of animals," because they have such names as Aries, the ram; Taurus, the bull; Cancer, the crab, etc. Actually, in modern times it is believed that the amusement park ride called the "merry-go-round" got its inspiration from the zodiac. Astrology is a blend of understanding the gravitation and magnetism of the earth's relationship to these heavenly bodies contained in our solar system and the Universe beyond. These forces are far greater than what meets the eye. They act like a symphony of invisible cross-currents and alignments that can affect the earth and the physical, emotional, mental and spiritual evolution of its inhabitants.

As we look at astrology through contemporary eyes, it is really not so implausible to regard planets as functioning in a manner similar to satellites produced by humans. Just as there are many types of satellites with different specialties and purposes such as communications, weather forecasting, spying, etc., so it is with the metaphysical properties and specialties of planets. Planetary energies trigger many forms of awareness in the human species, both on a collective and an individual basis. The trick is to use the right satellite for the right intention or purpose.

The Birth of Venus and the Rose

The birth of Venus in mythology is very unusual and telling. Venus was born out of conflict, from the violence between father and son, Uranus and Saturn. Her birth, according to myth, takes place when Uranus rejected his own creations in the form of his children by Gaia. As soon as each child, or creation, was born, he did not find them as perfect as his original concept of them, so he repeatedly shoved each child, as it was born, back into the womb of Gaia, Mother Earth. It is important to remember that this is esoteric language for an archetypal force springing out of chaos. It is deliberate in its attempt to represent that those we would call "gods" are not subject to the same rules of human procreation.

Weary of having her own children rejected by the 'creative force', represented by the Sky God Uranus, Gaia instructed their son Saturn, or Kronus in the original Greek, to await the approach of Uranus and castrate him. This is a deep allegory to demonstrate the consequences of thoughtless rejection of one's own creations, thus sacrificing the possibility of evolution to the pursuit of perfection.

The legend continues that Saturn threw the severed genitals of his father Uranus into the sea where a great and turbulent roiling of the waters began. The agitation of the waters increased to the point where a great foam was produced, and from this enormous foam, standing on a seashell, emerged the beautiful Aphrodite, which means "she who is born of the foam," or Venus, as we call her in Western astrology.[9] Among her many lovers were Mars, Hermes, and

Adonis. In a jealous rage, Mars transformed himself into a savage wild boar and gored Adonis to death. Roses sprang forth from her tears at the death transformation of her lover or from the actual drops of his blood as they fell, depending on which version one chooses. The myth varies from region to region throughout the vast Greek Empire. Henceforth, the rose, the symbol of Venus, like the birth of Venus herself, became associated with beauty being born of violent transformation.[10]

Venus, Phi and the Rose

The rose has always been associated with Venus in Western astrological mythology. Just as the rose is a symbol of serene beauty, so also is the planetary influence of Venus and the signs that it rules, Taurus and Libra. Wherever we see Venus in the astrological chart, either where in the chart it was at birth or by transit, or where it is located in the chart on a later date, we also associate the symbol of the rose. It points to areas of our lives where we can embrace healing and diplomacy. It shows us that Beauty can be the doorway to higher consciousness.

We can also look at the rose as an analogy for the relationship between Body, Soul, and Spirit.[11] If we look at the beginning of the growth of a rose, we see a hard shell or sheath, representing the physical body, which is supported by a stem, or cord, attaching it to the earth. Eventually, the petals of the rose burst forth from the covering of the bud and open into their full magnificence, still nurtured by the vessel and support of the stem that connects it to its source. The rose exerts and exudes a magnetism that attracts other life, such as bees and humans, to assist in its evolution. Eventually, the petals dry, fall away and yield to the production of the seed, representing the spirit, from which the cycle can once again bring forth beauty into the physical world. This is a constructive manner in which to view Venus in the natal or transiting birth chart.[12]

The spiral of a seashell, the wave in the ocean, the shape of a galaxy, the needles of a pine cone and the petals of a rose are but a few of countless examples that demonstrate a number system discovered in the middle ages by an Italian mathematician named Fibonacci, more properly called Leonardo of Pisa (ca. 1175 – ca. 1250). A contemporary of St. Francis of Assisi and Marco Polo, he was educated by the Moors in Northern Africa during his youth when Europe was awakening to the Age of Exploration. He introduced the decimal system to Europe and was one of the first to present the Hindu-Arabic number system to European scholars, thus playing a significant role in causing Europe to change from the old and unwieldy system of Roman numerals to Arabic numbers.[13] The interesting concept of the Fibonacci number system is its being based on the numerical relationship of the spiral rather than the circle. The system is really quite easy to understand. A Fibonacci number is always the result of two preceding numbers that are in natural sequence with each other. The basic system is 1, 1, 2, 3, 5, 8, 13, 21, etc. Note that "2" is a combination of 1+1, the "3" is a result of 1+2, the "5" is a combination of the preceding 2+3, the "8" is a sum of 3+5, etc.

This formula is called "phi" and relates to the spiral. Another mathematical formula that we are all taught in school is that of "pi," which relates to a circle and its diameter. Now, this is important: the formula "phi" relates to the "spiral of life," which represents evolution and enlightenment; the "pi" formula relates to the circle, which is a loop or infinite manifestation of the status quo.

In terms of astrological psychology, defined as the relationship of the human subconscious to planetary movement, every time a planet returns to where it was when a person was born, there is an opportunity to take an evolutionary step upward in awareness and consciousness. If the person does not progress in awareness during each planetary return, he is symbolically "going around in circles."

As Venus orbits the earth, it makes eight inferior conjunctions that occur when Venus is exactly between the Earth and the Sun every eight earth years, and if plotted on a graph, these conjunctions form a pentagram, a five pointed star, which is the symbol of the "Goddess." In terms of geometry, a pentagram is composed of five golden triangles. All of these triangles symbolically connect our perception of the material level of awareness with the cosmic or spiritual understanding of our relationship to the Universe.

Whenever Venus returns to the position in an astrological chart where it was at birth, a chart can be calculated for that moment to forecast the areas of life that will inter-relate with the current Venus cycle of the individual. There is a phi or Fibonacci relationship between the Earth and Venus in terms of their respective orbits. For example, the earth orbits the sun 8 times during the same time frame that Venus orbits the sun 13 times. 8 and 13 are phi, or spiral, evolutionary numbers.

Venus and Taurus

Venus rules the signs Taurus and Libra, and in astrological practice, the qualities attributed to a planet also apply to the zodiacal signs that it rules. Venus is inextricably associated with the rose, and since Venus rules the sign of Taurus, the rose is also associated with Taurus. The part of the body associated with Taurus is the throat and therefore the voice and the fifth chakra. People with prominent Venus, or with many planets in Taurus, or those whose rising sign is Taurus usually have very pleasant voices and speak in well-modulated phrases. There is solidity to the speech and a very grounded style of communication.

Venus rules the second house of personal values in the natal chart. People with Venus or Taurus located in this area resonate with quality and durability. It is interesting that Venus, in astronomical terms, rotates on its axis from west to east instead of from east to west like most of the planets. Therefore, if we were living on the planet Venus, the sun would rise in the west and set in the east. This reverse direction of axis rotation causes Venus to radiate strong magnetism. Those with prominent Venus energy in the chart usually draw things to them in terms of resources and helpful people. They collect and amass possessions and people and have great difficulty throwing things away.

The Rose and its Connotations in the East and the West

The rose can bear a similar description. It is a flower of quality and, while on the vine, longevity. Roses when picked for a bouquet usually require a more formal container than most other flowers. While each flower is perfection in itself, its quality can stand alone in a single bud vase and inspire great appreciation and contemplation. Also, since people with planets in Taurus love abundance, they present roses, often in large bouquets by the dozens!

Being always mindful that Venus is associated with the throat chakra, we come across a more profound association with the potential of the rose. In Kabbalistic or Hebrew astrology, this fifth chakra is called "Da'ath" and is considered the doorway to the Underworld where both psychological and spiritual evolutionary transformation takes place. Hebrew astrology considers this Da'ath to be the energy point on the human body that serves as an initiation chamber to gain the wisdom to determine the difference between spiritual darkness and light, a process that can seem terrifying. The rose is connected both with romance and Eros. Remember, the rose has thorns! They teach us that the rose must be approached with respect if we are to fully hold and experience the gift of its beauty.

The symbolic meaning of the rose in Western civilization is equivalent to that assigned to the lotus blossom in ancient Egypt, India, and the Far East. It is the symbol of the elevated level of consciousness of the human who has awakened to his true identity. The fragrance of the rose has been long prized in many cultures, especially those of Persia and Bulgaria. These cultures are closely identified with essential rose attar or fragrant oils. To make one ounce of rose essence, over sixty pounds of rose petals are distilled. The rose is a prize and one of the hallmarks of idealized beauty.

Venus and Libra

The sign of Libra is the other domain or sign ruled by the planet Venus. While the aspect of Venus that we see in Taurus is sensual, acquisitive, and erotic, Venus in Libra represents the diplomat, the keeper of peace, the patron of fine arts. Remember that Venus represents beauty born of conflict as she rose out of the sea, standing on a seashell. Incidentally, the wave and the shell are two examples of spiral numerology beautifully immortalized in the famous painting by Botticelli. Here in Libra, the planet is concerned with the conflict of balancing relationships, whether between nations or individuals. It is no co-incidence that sending roses to promote romance and harmony is a long established tradition, especially on Valentine's Day.

Actually, Valentine's Day has its roots in very ancient Greco-Roman times when the celebration of their marriage honored Juno and Zeus in early February. The occasion featured festivities in the name of the goddess Venus, her son Eros and the magic of romance. Later, during the reign of the tyrannical Roman emperor Claudius, tradition dictated that young men and women lead lives very separate from one another. But during the festival of Juno, the young maidens would put their names in a jar, and the young men would choose the name of a maiden who would then be his significant other for the following year. If they fell in love, they would marry; if not, there would be another year and another name to draw. Claudius outlawed this celebration, but a bishop, St. Valentine, would continue to secretly marry the couples against the emperor's wishes. Claudius had St. Valentine beheaded on February 14th around the year 270. After the death of Claudius, the festival was reinstated, and during these old pagan days with worship of Venus and Eros, the young men would once again choose a name from a jar and pick a young maiden to court for the following year. The maiden chosen was afterward known as his "Valentine."

The Rose and the Divine Feminine

From earliest times, Venus has been revered as one aspect of the Divine Feminine in religious and spiritual practices. During the era of Christianity's initial spread throughout Europe, a representative of the Feminine was necessary to balance the severity of the masculine aspect of the faith. The preceding cultures and religions were very oriented towards honoring the Goddess energy in their rituals and symbols. Hence, the increased veneration of the Virgin Mary and Mary Magdalene established itself in Christian theology of the time. The rose was not allowed to represent the Virgin Mary because of its long association with Venus, a Goddess of pagan times. Therefore, the lily was associated with the Virgin Mary because of its connotation with the concept of purity. The rose became associated with Mary Magdalene because of her relationship to sensuality, a basic quality associated with Venus. This is evidenced in the magnificent stained glass windows and cathedral sculpture throughout medieval Europe. Prime examples of this are the vividly colored "rose" windows in such major cathedrals as Notre Dame de Paris and Chartres. Many of those cathedrals were erected on the original sites of Goddess worship, which also served as ancient astrological observatories. There, particular attention was paid to the dual qualities of Venus: because of orbital synchronicities, sometimes Venus is seen as the "morning star" and sometimes as the "evening star."

Conclusion

In the language of both sacred and cosmic geometry, the rose remains a symbol of beauty, combining the numerology of the macrocosmic and the microcosmic universe of the golden spiral of phi discovered by Fibonacci. Just as the meditative contemplation of the form, structure and substance of the rose can lead to a greater reflective and parallel understanding of ourselves, so by harmonizing with those planetary transits that eternally work their cosmic effect on the human condition might we also open our understanding of unfolding growth. We have only to awaken to the lesson of the rose petals' pattern to follow the spiral of evolution instead of the circle of limitation.

Today, as human understanding of the art and science of astrology explodes with new awareness, ancient myths become clearer to us. The balance necessary to maintain an evolving insight into creating our lives, rather than merely administrating life events into some understandable pattern, rests in comprehending the sacred symbols of the distant past. The magic of today is the science of tomorrow.

The rose is prevalent in every astrological myth that involves appreciation for the vessel of feminine beauty. Perhaps it is best summed up by a little memorial plaque that graces a bench in New York's Central Park: "Myths are stories that never happened, but always are."

References

[1] Ariel Guttman and Kenneth Johnson, *Mythic Astrology* (St. Paul, Minnesota: Llewellyn Publications, 1993) 2.

[2] Ibid. 3.

[3] Robert Hieronimus, Ph.D. *America's Secret Destiny: Spiritual Vision and the Founding of a Nation.* (Rochester, VT: Destiny Books, 1998) 37, 44.

[4] Ariel Guttman and Kenneth Johnson, *Mythic Astrology.* 1.

[5] James Redfield, *The Celestine Vision* (NY: Warner Books, 1997) xix.

[6] Ibid. xxiii.

[7] Ibid.

[8] C.L. Barnhart, ed., *The American College Dictionary* (New York: Random House, Inc., 1962) 65.

[9] Ariel Guttman and Kenneth Johnson, *Mythic Astrology.* 41-51.

[10] Ibid. 51.

[11] Lisa Tenzin-Dolma, *Understanding the Planetary Myths* (London: Quantum, 2005), 53.

[12] Ibid. 56-57.

[13] Richard A. Dunlop, *The Golden Ratio and Fibonacci Numbers* (Singapore: World Scientific Publishing Co. Pte. Ltd., 1998).

7

ROSES IN TURKISH LITERATURE AND CULTURE[1]

By Gamze Demirel[2]

For centuries accepted as meritorious and holy, nature, mirroring God, has been a source of inspiration to artists. The Turkish people revere nature and especially the flower known as the rose. Writing in 2004 in a phenomenological approach to the study of Islam, Annemarie Schimmel saw the rose as the most precious of garden flowers in Islamic culture and the "sultan" of the genre. "Especially the red rose," she said, "indicates Divine Greatness" (39).

The rose has a remarkable place in the Islamic world of imagery as narrated in the *Qur'an,* the Holy Scripture of Islam, and secondary Islamic discourse. The story of the Prophet Abraham, for instance, has given birth to the rose-fire-Abraham trio in classical Turkish poetry. Prophet Abraham, one of the Qur'an's greatest seers, also plays a significant role in the Christian *Holy Bible,* but his story begins when God orders him to leave his land for an unknown place that God will show him (Hebrews 11:8). The years of Abraham's youth are elaborated in the *Qur'an* in chapters 2:258–260, 6:75–84, 19:41–50, 21:51–70, 26:69–104, 29:16–25, 37:83–98, and 43:26–27. According to the Qur'anic account and further elaborations of the Prophet Muhammad, Abraham's father was an idol-maker for a self-proclaimed divine king. After Abraham demolished all the idols in the shrine, the king decided to punish him by throwing him into a huge fire. But instead of burning the victim, the flames turned into a rose garden. In classical Turkish poetry, this image of Abraham, together with the Prophet Joseph's story, provides source material for numerous allegories of the rose.

Joseph, another prophet mentioned in the Qur'an, is the subject of a coherent narrative told in a chapter that bears his Arabic name, Yûsuf (The *Qur'an,* Chapter 12). The figure corresponds to the Biblical Joseph, the eleventh son of Jacob (Ya'qub). Believed to have been a very handsome and attractive boy, he is the smartest among his brothers, so clever in fact that jealousy caused them to throw him into a well. Rescued by a passing caravan, he was brought to Egypt to be sold as a slave to the King's finance minister (Potiphar in the Biblical account). When Joseph matured, Zuleikha (Potiphar's wife) fell in love with him and wanted to seduce him. His resistance lead her to tear his shirt from behind. In Classical Turkish poetry, this incident is likened to the blossoming of the rose; we see the emotional assault as a tearing off of petals.

Another Qur'anic and Biblical figure associated with roses in Classical Turkish poetry is the Prophet Solomon. In Islam, it is believed that, God willing, the wind would provide this king's transportation. Needing to travel, Solomon would simply mount his throne, command the wind, and be carried even to the farthest distance in the realm. In Classical Turkish poetry, his throne is likened to rose petals flying in the air with the blowing wind. And the shape of the ring that he used to govern birds, djinns, and all creatures is likened to the rosebud. The blooming of the rose is also used as a metaphor for the resurrection of the dead when Israfel (in Hebrew Raphael) announced the coming of the end of the world by blowing the trumpet called Sûr the second time. It is believed that his first blow would bring the Apocalypse and his second start resurrection in the Hereafter.

In Turkish Islamic tradition, however, the rose represents most power-fully the Prophet Muhammad. Indeed, the Prophet Mohammed even sweats the scent of a rose; therefore, Muslim Turks hold, if you sniff a rose, you have done a good deed, an attitude that helps explain the intimate love of Turkish people for nature and especially for flowers. Many Turks believe that each flower has its own language to help explain a person's loves and sorrows. "One can discover leaves of dried flowers like roses and violas while thumbing through the pages of old books" (Onay, 2000: 221). These relics, possible gifts to express love and interest, had then been hidden away as souvenirs.

Not only blooms but also by-products serve as cultural coin. Tradition-ally, Turks put rose oils into flasks for traveling. Turks serve rose water to guests in inlaid cases called *gülabdan*, and they wash mosques with rose water. A person's blush is likened to the color of a rose, emphasizing that the rose, as well as decency and shame, has profound significance in Turkish culture.

The name of the rose and its derivatives have titled many books publish-ed in Turkey about Turkish culture. For example (*Gülbün* [Rosary], *Gülşen* [Bed of Roses], *Gülzar* [Rose Garden], *Gül-deste* [Bunch of Roses] etc.). The ceremonial cardigans Mawlawi and Bektashi[3] people used to wear for rituals were called "Bunch of Roses." Gravestones, too, were embellished by patterns of the rose. The rose symbolism occurs in many works of Turkish literature and art, often standing for the palace. As Kortantamer (1993) writes in her analysis of Old Turkish literature, "the 'rose' makes its whereabouts a center, a sort of palace, thanks to its shining like the sun" (413).

Because the rose blossoms in spring and brings beauty, color and happiness to gardens, another name of the spring in Turkish culture is "the season of the rose":

> *Nevbahâr oldu gelin azm-i gülistân edelim*
> *Açalım gonce-i kalbi gül-i handân edelim*
> (Bâkî 4- G5. 323, b. 1) (in Küçük, 299)
> (In English: Now the spring has come. Let's go to the rose garden, open the buds of our hearts and create a smiling rose [that is, a rose in bloom].) (Yesirgil, 1963: 57)

> *Gül devri ayş eyyâmıdır zevk u safâ hengâmıdır*
> *Âşıkların bayramıdır bu mevsim-i ferhunde-dem*
> (Nef'î 6- K7. 15, b. 3) (in Akkuş, 94)
> (In English: Time of the Rose, spring in other words, is for feasting, gallivanting, living on the primrose path; this season, an era of happiness, is the holiday of lovers.)

Turkish families who give considerable thought to naming their children often call them "gül," that is, "rose" and names evoking roses (like *Gül, Gülcan, Gülben, Güllü, Gülsever, Goncagül, Sadegül, Esengül, Badegül, Birgül, Şengül, Songül* etc. (especially but not only girls). Similarly, many mosques, districts, streets and avenues have such names – clear evidence of the importance granted to the rose in Turkish culture.

Moreover, numerous proverbs and idioms include the word "rose" in our language: *Dikensiz gül olmaz* (There is no rose without its thorn),[8] *El bebek gül bebek* (blue-eyed boy),[9] *gül gibi bakmak* (look after like a rose),[10] *gül gibi geçinmek* (get along well),[11] etc.

As a generic, "gül" in Turkish can mean all flowers, as seen in certain modes of Classical Turkish music: "for instance Gonca-i Rânâ, Güldeste, Gülizâr, Gülşen-i Vefâ, Gülzâr, Gülruh" (Ayvazoğlu, 1995: 100). Many classical compositions include the word "rose" and its derivatives. Rosebud *(gonca gül)* expressed in the following lines means lover, liver *(ciğer)* means heart, and break in heart *(yâre)* means heartache: *"Bir gonca gülün yâresi vardır ciğerimde"* (Dede Efendi).12 (Literally, I have a broken rosebud in my liver.)

The twentieth-century Turkish historian Reşat Ekrem Koçu (1905 - 1975) says in one of his soulful articles titled "Rose Oil" that the best ballads of Rumelia were sung while roses were boiling (in Ayvazoğlu, 1995: 198). Here is a ballad:

> *"Ellerinde gül oya/ Gülmedim doya doya*
> [In English: Rose lace in your hands/I have never laughed to my heart's content] (in Ayvazoğlu, 1995: 184).

Furthermore, the hymns and blessings chanted loudly that had been prevalent among Janissaries, Kalenderis (a group of Dervishes), Bektashis and Mevlevis were called *Gülbang. Gülgeşt* evokes wandering in rose gardens in those days. The circular ornamental metal or wooden plate with a hole in the middle for a door knocker is also called *ağızlık gülü* (nozzle rose) (*Encyclopedia of Islam*, 1996: "The Rose").

The rose, gesturing toward the other realm (that invisible, unknown, secret world) by means of its color, aroma and shape, has also been used to explain mortality in earthly life. Life is short, almost too short for growing roses:

> *Bir zebân-ı hâldür her yaprağı fehm itseler*
> *Perde-dâr-ı hâk olanlardan virür ahbâr gül*
> (Fuzûlî 13 - K. 9, b. 33) (in Akyüz, et al., 46)
> (In English: Each living rose petal suggests others buried under the soil – the dead – each having lived and died in its own distinct way.)

As to the rose's expressive utility, in olden times when directly revealing such wishes to parents would have been deemed disrespectful, Turkish youth announced their desire to marry by using various symbols. For instance, a girl or boy donned "socks with roses" to tell the family he or she had reached marriageable age. Similarly, it is common tradition to swaddle babies with rose-patterned clothes. "It is also known that rose petals were scattered on beds in the past" (Onay, 2000: 221). In addition, roses figure significantly in Turkish handicrafts, rose-shaped patterns (rose lace), for instance, or a heavy, silky type of fabric called *"Güllü Dîbâ"* imprinted with the flower (*Encyclopedia of Islam*, 1996: "The Rose").

Furthermore, throughout Turkey rose rhymes and references are used in the names of protagonists of famous anonymous love poems such as *Varka ile Gülşah* (Varka and Gülşah), and *Hüsrev ile Gülruh* (Hüsrev and Gülruh), which were written in the 14th century masnawi style (consisting of an indefinite number of couplets, with the rhyme scheme aa/bb/cc). As in many other cultures, "6th of May is accepted as *Hıdırellez* (celebration of spring) in Turkish culture. In *Hıdırellez*, people leave notes containing their desires under rose trees where prophets Hızır (an immortal person thought to come in time of need) and İlyas (Elijah) are believed to have met and throw these notes into the sea, to the kingdom of Hızır, while the sun is rising" (Gezgin, 2010: 82). Another source states that "… the oil roses blossom in May and are boiled in the boilers laid on the 6th of May, namely *Hıdırellez*, till the end of May" (Ayvazoğlu, 1995: 198).

Rose oil and rose water were not merely aromatic and therefore beautiful but also useful: they belonged to the inventory of physicians in the Ottoman period beginning at the end of the 13th century. Sources confirm that roses in the gardens of Sufi monasteries provided medical benefits:

> *Bulunur her derde istersen gülistânda devâ*
> *Hokkasında goncenün san kim şifâ cüllabı var*
> (Fuzûlî- G. 74, b. 4) (in Akyüz, et al., 167)
> (In English: If you wish, you can find a cure for all your problems in the rose garden, in curative rose water in the pot of the bud.)

If diarrhea is serious, patients were advised to "apply to the stomach a poultice made of the pulp of rose, celery, anise and cumin seeds; or of cumin seeds and rose seeds in vinegar; or oat seeds boiled in vinegar" (Kahya, 1995: 236).

Sometimes, children's digestion weakens. This illness is cured by applying rose water, lily and myrtle (ibid. 240). For instance, when saffron is mixed with rose, camphor and coral, it penetrates into the heart (ibid. 38).

Should these mixtures fail and health fade, the rose at least goes on: you find in Anatolia innumerable tombs named *Gül Baba* (Father Rose) expressing the flower's symbolic meaning in Turkish Sufi thinking. The rosebud refers to unity (union, God) and the rose in bloom to abundance (wealth, everybody and everything on earth, as well as all that prevents human beings from reaching their creator).

Jam and syrup are made from roses. Desserts like *güllaç* (starch wafers filled with walnuts soaked in milk and rose water), *su muhallebisi* (starch pudding with rose water), *güllabiye* (halva with almond and rose water and garnished with rose petals) were the indispensable flavors of Ottoman cuisine.

The rose has had an important place in the daily lives of Turkish people from the Ottoman period till today. Commonly used in cosmetics (perfume, soap, etc.) the flower plays an important role in generating Turkey's agricultural income as well.

The rose – a generic for all flowers – is also a common symbol of love as well as a sign of mystery and sanctity. The legendary love between the rose and the nightingale is inexhaustible in Turkish writing, as in world literature in general. Its color, smell, shape, thorns, brief lifespan and suggestive buds have made it a recurring trope in classical poetry. The classical Turkish poem best known in the West is the masnawi by 17th-century Ottoman poet Kara Fazli called *Gül ü Bülbül* (The Rose and The Nightingale). Many Ottoman poets wrote odes with the "rose" rhyme including Necâtî (15th century) and Nev'î (16th century). But among these odes, one of the best is Fuzûlî's written for Suleiman the Magnificent (1495-1566), the tenth Ottoman Sultan, under the penname Muhibbî. The stanzas' rich image world offers a different play on words in nearly every use of the term "rose" in sixty-two couplets (*Encyclopedia of Islam*, 1996: "The Rose"). And in the following couplet, the 18th century-Ottoman poet Nedim loves the rose as a form of addressing his beloved:

> *Gülüm şöyle gülüm böyle demektir yâre mu'tâdım*
> *Seni ey gül sever cânım ki cânâna hitâbımsın*
> (Nedim- K. 19, b. 6) (in Macit, 74)
> (In English: It is my habit to call the beloved "my rose," so I love the rose for giving me a name for my beloved.)

In spite of cruelty, arrogance or disloyalty, the lover welcomes the beloved as a rose, an attitude also reflected in the tradition of fixing a rose on the head or turban (*Encyclopedia of Islam*, 1996: "The Rose"):

> *Öyle pinhân eylemiş gögsinde sırr-ı ışkunı*
> *Kim ayağından asarlar eylemez izhâr gül*
> (Fuzûlî- K.9, b. 24) (in Akyüz, et al., 45)
> (In English: The rose keeps the secret of your love in its heart and cannot give it away for fear of being hanged by the foot, that is, being pinned upside-down on the turban or chest.)

Furthermore, lovers' tears are rose-colored (red) or in sprinkled droplets of rose water:

Urunca şâne gîsû-yı hayâl-i yâra müjgânım
Gülâb-efşân olur yâd-i ruhıyla çeşm-i giryânım
(Nef'î- Müfred14 s 1) (in Akkuş, 352)
(In English: Envisioning my beloved, my crying eyes sprinkle rose water when I recall her face.)

The lover can't take his eyes off the cheek of the lover that resembles a rose:

"Günde yüzün elli kerre görsem de,
Tamahkârım, gözüm doymaz gül yüze"
(in Özalp, 468)
(In English: Even seeing the face of the beloved fifty times a day wouldn't satisfy me, for I am greedy. I cannot be happy without those rosy cheeks. In other words, the lover wants to remain always in the presence of his beloved.)

We might also say that this lover is enflamed, associating roses and fires, a plausible interpretation since classical Turkish literature often reminds us that whatever stands upright overcoming obstacles is a vertical element, and these vertical elements are celestial and divine. Both roses and fires partake of this association.

The similarity of color between roses and fires has been the source of many images in Turkish poetry:

Gül âteş gülbün âteş cûy bâr âteş
Semender-tıynetân-ı aşka bestir lâlezâr âteş
(Şeyh Galip15 - G. 139, b. 1) (in Kalkışım, 327)
(In English: The rose is fire, the rose sapling is fire, the rose garden is fire, and the river is fire. Fire is enough for the salamander-like lovers in a tulip garden/ The rose, the rose sapling, the river turning red in the evening and the poppy garden share their color with the blaze.)

The rose pattern commonly appeared in post-Tanzimat[16] Turkish literature, written in a period known for accelerated Westernization in all aspects of the declining Ottoman Empire. Thus, despite considerable change and novelty, the rose remained an indispensable literary trope. Recaizade Mahmut Ekrem (1847-1914) expresses his love for the rose and the nightingale in his book *Pejmürde* (Dishevel), while nationalist poet and writer Namık Kemal (1840-1888) uses the rose as the symbol of nation and flag in his poem "Vaveyla":

> *Âh böyle gezer mi hiç canan?*
> *Gül değil arkasında kanlı kefen...*
> *Sen misin, sen misin garîb vatan?*
> (Namık Kemal, Vaveyla, Nefta 1)17
> (In English: Alas, does the beloved ever go around like that?/Behind her is not a rose, but a bloody shroud/ Is that you, is that you? Oh lonely homeland!)

The rose maintains its irrevocable fame in Turkish literature in the Republican Era that started with the establishment of the Republic of Turkey as a successor to the fallen Ottoman Empire in 1923. The Second New *(İkinci Yeni)* current in Turkish poetry that was shaped in the 1950s by poets like Edip Cansever, İlhan Berk, Cemal Süreya, Sezai Karakoç and Ege Ayhan frequently used rose metaphors to bring esoteric language back to Turkish poetry which had been de-popularized by the "Garip" literary trend, a movement to nationalize the Turkish language once the modern nation-state had taken root.

Therefore, Second New current poets like Behçet Necatigil (1916-1979) and Ahmet Muhip Dranas (1909-1980) challenged conventional patterns in language, forcing the limits of word ordering by changing or disrupting idioms. Emphasizing imagination and feelings in poetry, they privileged abstract metaphors and made the rose pattern an essential poetic tool.

Ahmet Haşim (1884-1933) was among the pioneers of symbolism in Turkish literature, even before the emergence of the Second New current in the 1950s. The rose is the flower he loves most, especially the image of "the bleeding rose" in "Merdiven" (The Stairs), a remarkable poem:

> *"Eğilmiş arza, kanar, muttasıl kanar güller;*
> *Durur alev gibi dallarda kanlı bülbüller,*
> *Sular mı yandı? Neden tunca benziyor mermer?"*
> (Ahmet Haşim/ Merdiven)
> (In English: Bent towards the ground, the roses are constantly bleeding;

The bloody nightingales stand on fire-like branches;
Have the waters been burnt? Why does the marble look like
bronze?)

Another influential poet and writer of the early Republican era was
Yahya Kemal Beyatlı (1884-1958), who mentioned the rose on various occasions in his poems, especially in "Rindlerin Ölümü" (Death of the
Dervish) and "Rindlerin Akşamı" (Night of the Dervish), two of the most
important works presenting the "rose" pattern in an impressive way.

Similarly, in "Endülüs'te Raks" (Dance in Andalusia) the rose gives its
color to the entire piece:

> *Zil, şal ve gül*
> *Bu bahçede raksın bütün hızı...*
> *Şevk akşamında Endülüs üç def'a kırmızı ...*
> ...
> *Gül tenli, kor dudaklı, kömür gözlü, sürmeli ...*
> *Şeytan diyor ki sarmalı, yüz kerre öpmeli..*
> *Gözler kamaştıran şala, meftûn eden güle,*
> *Her kalbi dolduran zile, her sineden*
> (In English: The bell, the shawl, and the rose
> All the speed of dance is in this garden
> Andalusia is three times red on this night of pleasure
> ...
> Rose-skinned, ember-lipped, coal-eyed blackened with kohl
> Satan says one should hug and kiss her a hundred times,
> To the splendiferous shawl, to the enchanting rose,
> From all bosoms to the bells filling all hearts)

Rindlerin Ölümü / Death of the Dervishes

> *Hafız'ın kabri olan bahçede bir gül varmış;*
> *Yeniden her gün açarmış kanayan rengiyle.*
> *Gece; bülbül ağaran vakte kadar ağlarmış*
> *Eski Şiraz'ı hayal ettiren ahengiyle.*
> *Ölüm asude bahar ülkesidir bir rinde;*
> *Gönlü her yerde buhurdan gibi yıllarca tüter.*
> *Ve serin serviler altında kalan kabrinde*
> *Her seher bir gül açar; her gece bir bülbül öter*

(Yahya Kemal Beyatlı, in Ayvazoğlu, 238)
(In English: Death of the Dervishes
There was a rose in the garden where Hafiz was buried
Each day it bloomed again with its bleeding color
At night, the nightingale wept till dawn
With its melody reminiscent of Ancient Shiraz
Death is a peaceful land of spring for the dervish
His heart fumes like a censer for years
And in his grave under cool cypresses
A rose blooms each morning; a nightingale sings each night.)

Rindlerin Akşamı

Dönülmez akşamın ufkundayız, vakit çok geç
Bu son fasıldır ey ömrüm nasıl geçersen geç
Cihana bir daha gelmek hayal edilse bile
Avunmak istemeyiz öyle bir teselliyle
Geniş kanatları boşlukta simsiyah açılan
Ve arkasında güneş doğmayan büyük kapıdan
Geçince başlayacak bitmeyen sükunlu gece
Guruba karşı bu son bahçelerde, keyfince
Ya şevk içinde harab ol, ya aşk içinde gönül
Ya lale açmalıdır göğsümüzde yahud gül
(Yahya Kemal Beyatlı18)
(In English: Evening of the Dervishes
We are at the end of a no-return path [evening], too late it is
This is the last episode, oh my life,
may you pass however you like
Even if coming back to the world once again is dreamed of
We do not want to be consoled with such a solace
When you pass through the grand gate
with its widely opening wings in dark space
And where no sun rises behind it
Then will start the night with unending silence.
Against the sunset in the garden of life, as you like
My heart, be ruined either in desire or love
In your bosom, should bloom either tulip or rose.)

Thus, the rose remains the beloved flower of Turkish literature with all its connotations. Sezai Karakoç (1933-), a contemporary Turkish poet, thinker, and politician, heralds the coming of a civilization shaped by Prophet Muhammad associated with the rose in *Gül Muştusu (The Gospel of the Rose)* (1974). The rose symbol is also the main topic of *Güller Kitabı (The Book of Roses* republished in 1992). This important work, written originally by the Turkish historian Beşir Ayvazoğlu (1953-) traces the role of flowers, especially the rose, in Turkish culture. He analyzes the feelings of Turks toward flowers and nature in general from nomadic tribes to great civilizations. Only in Nazım Hikmet Ran (1902-1963), a famous Turkish poet from recent times, do these feelings towards the rose seem negative in contrast to the tradition that had come before:

> "Güle, bülbüle, ruha, mehtaba, falan filan/ karnımız tok/
> Ve şimdilik/ gönül işlerine vermiyoruz metelik."
> (English: We don't believe in roses, nightingales, spirits, or moon-
> light /and for now/we have no regard for affairs of the heart)

In sum, the rose has always attracted great attention as an important element not only in Turkish but also in world culture and literature as confirmed by innumerable legends and beliefs in both East and West. Turks in particular have integrated roses into every aspect of their lives. Indeed, Turkish culture has been shaped by the love of the rose in all its beauty and meaning. With its rich accretion of symbols extending from Adam and Eve and the Prophet Abraham to the Prophet Mohammed, the rose has a cultural background as old as the history of humanity. We might even say that, in literature, architecture, poetry, music, Sufism, calligraphy, illumination, engraving, house decorations, gravestones, ceramics, mural paintings, clothes, names, jewelry, behaviors, value judgments, thought systems and lifestyles, the rose is emblematic of the Turkish nation itself.

Notes

[1] I would like to give my special thanks to Semiha Topal for help translating this article into English.

[2] Assistant Professor of Classical Turkish Literature in Süleyman Şah University, Faculty of Humanities and Social Sciences, Department of Turkish Language and Literature, Istanbul, Turkey.

[3] Mawlawi (Mevlevi in Turkish) is a Sufi order founded in 1273 in Konya, Turkey, by followers of the great Sufi Mawlana Jalal Ad-Din Al-Rumî, known commonly as Rumi in the West. They are known as the Whirling Dervishes due to the special zhikr ritual named

Sema. The Bektashi order was also founded in Turkey within the same time period (13th century) by the followers of Hadji Baktash Wali (commonly known as Bektaşi in Turkish).

[4] Bakî was the pen name of the Ottoman Turkish poet Mahmud Abdülbâkî (1526-1600 AD).

[5] "G" is the abbreviation of "gazel (ode)," one type of poem in Classical Turkish Literature and "b" is the abbreviation of "beyit (couplet)," one of the poetry units that combines two lines.

[6] Nef'î (1572-1635) was an Ottoman poet and satirist most famous for his odes (kaside).

[7] "K" abbreviates "kaside (ode)," one type of poem in Classical Turkish Literature.

[8] Every good and beautiful thing has a disruptive aspect (Proverb).

[9] Spoilt and sassy (Idiom).

[10] 1) Making a living without financial difficulty; 2) Looking after oneself in a good and clean way (Idiom)

[11] 1) Getting along really well and keeping in good with someone; 2) Living comfortably without vast opportunities.

[12] Dede Efendi (1778-1846): His full name is Hammamizade İsmail Dede Efendi, a well-known composer of Turkish classical music.

[13] Fuzûlî (c. 1483-1556) was the pen name of the Ottoman poet, thinker, and writer Muhammad Bin Suleyman. He wrote odes (kaside) in Azeri, Persian, and Arabic.

[14] Müfred is one type of Classical Turkish poetry.

[15] Şeyh Galip (1757-1799): A Classical Turkish poet who was also a Sufi.

[16] The Tanzimat, meaning reorganization of the Ottoman State, was a period of reformation that began in 1839 with the declaration of the imperial edict aiming to modernize the Ottoman State, and to integrate non-Muslims and non-Turks more thoroughly into Ottoman society by enhancing their civil liberties. The period ended with the First Constitutional Era in 1876.

[17] See http://www.antoloji.com/siir/siir/gunun_siiri.asp?year=2011&month=7&day=15. Retrieved 16 January 2015.

[18] See http://www.siirperisi.net/siir.asp?siir=716. Retrieved 16 January 2015.

References

Akkuş, Metin. (1993) *Nef'î Dîvânı* (Collected Poems). Ankara: Akçağ Publishing.

Akyüz, Kenan, et al. (2000) *Fuzûlî Dîvânı* (Collected Poems). Ankara: Akçağ Publishing.

Ayvazoğlu, Beşir. (1995) *Güller Kitabı* (Book of Roses). Istanbul: Ötüken Publishing.

Gezgin, Deniz. (2010) *Bitki Mitosları* (Herb Mythos). Istanbul: Sel Publishing.

İbn-i Sina. (1995) *El-Kânûn Fî't-Tıbb.* Trans. Esin Kahya. Ankara: AKM.

Kurnaz, Cemal. *İslam Ansiklopedisi* (Encyclopedia of Islam). (1996) "Gül" (The Rose), İstanbul: T.D.V. Publishing.

Kalkışım, Muhsin. (1994) *Şeyh Galib Dîvânı* (Collected Poems). Ankara: Akçağ Publishing.

Kaplan, Mahmut. (1996) *Neşâtî Dîvânı* (Collected Poems). İzmir: Akademi Bookhouse.

Kortantamer, Tunca. (1993) *Eski Türk Edebiyatı: Makaleler* (Old Turkish Literature: Articles). Ankara: Akçağ Publishing.

Küçük, Sabahattin. (2002) *Bâkî ve Dîvânı'ndan Seçmeler* (Bâkî and Selections from His Collected Poems). Ankara: Ministry of Culture.

Macit, Muhsin. (1997) *Nedîm Dîvânı* (Collected Poems). Ankara: Akçağ Publishing.

Mazıoğlu, Hasibe. (1986) *Fuzûlî ve Türkçe Dîvânı' ndan Seçmeler* (Fuzûlî and Selections from His Collected Turkish Poems). Ministry of Culture, A Thousand Essential Works, Ankara: Ministry of Culture.

Karahan, Abdulkadir. (1985) *Nef' î Dîvânı' ndan Seçmeler* (Selections from Nef'î's Collected Poems), Ankara: Ministry of Culture.

Nevzat Yesirgil. (1963) *Bâkî: Hayatı, Sanatı, Şiirleri* (Bâkî: His Life, His Art, His Poems). İstanbul: Varlık Publishing.

Onay, Ahmet Talât. (2000). *Eski Türk Edebiyatında Mazmunlar ve İzahı* (Imageries and Their Interpretation in Old Turkish Literature). Ankara: Akçağ Publishing.

Özalp, Nazmi. (2000) *Türk Musikisi Tarihi-II* (The History of Classical Turkish Music – II). Ankara: Ministry of National Education.

Schimmel, Annemarie. (2004) *Tanrının Yeryüzündeki İşaretleri: İslama Görüngübilimsel Yaklaşım* (Deciphering the *Signs of God*: A Phenomenological Approach to Islam). Trans. Ekrem Demirli. İstanbul: Kabalcı.

8

The Rose in South India's Tamil Tradition

Uma M. Swaminathan

One of the most intriguing histories of the rose emerges from indigenous Tamil culture of Southern India and its prodigious love affair with roses. Growing up in India, I cherish having experienced the ancient wisdom passed down through generations in documented and oral form that traces in riveting texts the enigmatic role of roses over centuries. My interest in plants, herbs and spices, and their usage in different cultures had taken me back to the sub-continent a number of times to further this knowledge and research and to analyze how the ancient secrets, thousands of years old, can continue to benefit humanity.

The Rose and the Cosmic Cycle

Tamil culture had practiced the art of living through good hygiene, yoga and healthy foods all bound up with spirituality and the cosmic cycle. Here, the cosmic cycle refers to astrology, the time of day and movement of the moon and planets. According to treatises on Ayurveda, the ancient medical system of India,[1] the gravitational force of these celestial bodies affects the emotional patterns and physiology of the human body. For instance, the moon attracts water in the form of high tide and low tide. This has a bearing on humans whose bodies contain anywhere from sixty to eighty percent water depending on age and health. Therefore, to detoxify, fasting was recommended on full and new moon days. Breaking a fast with healthy foods was equally important. Keeping this in mind, sages found that a preventive technique using herbs, spices and fragrant flowers promised longevity, and bliss could be fostered, notwithstanding that a daily ritual had already made ample use of these ingredients, foremost among them being the rose.

The rose in all its glory, increasingly regarded as an elixir of life, became part of a tradition that ensured resilience and longevity. The rose taken internally is cooling, and externally its scent is soothing to the nerves. The rose adorns deities in temples and decorates alters in the privacy of homes. Womenfolk titivate themselves with rose flowers and rose essence. Rituals call for the use of fragrant roses, rose water and rose incense.

A traveler, the ambassador to the King of Persia, Abdur Razzak[2] who visited the southern kingdom of Vijayanagaram in 1443 writes, "Roses are sold everywhere. These people could not live without roses and they look upon them as quite as necessary as food." In fact, in various forms, roses were eaten.

Then and now, the rose is used as a tonic, an aphrodisiac and a digestive aid. A concoction of rose petals and honey called "gulkund" which is essentially a rose jam or rose preserve has been in existence since ancient times. It is enjoyed along with betel leaves called "Vethalai" in Tamil. Taken after a meal for proper digestion, it answers the viewpoint in the Ayurveda that all diseases are caused by indigestion and constipation. Moreover, Gulkund is dispensed along with other herbs to treat mental ailments and serves as a rejuvenating skin tonic. Rose hips are a store house for Vitamin C and are deployed as well for stomach disorders.

The Rose in the Vedic tradition

The source of religion and culture in India, Vedas, ancient Hindu scriptures in Sanskrit, are a set of ancient doctrines or Dharmic principles for mortals to follow. Among them the Rig Veda contains the most famous hymn on "the healing plants."[3] From ancient times, Ayurveda, the oldest of these medical systems, has made the rose an integral part of society.

For instance, the Ayurveda medical treatise containing prescriptions by Charaka Samhita, Susruta Samhita, Kashyapa Samhita and Bhava Prakasham Samhita imparted curative knowledge in terms of herbs and the rose in oral form; they provided a written record on palm leaf manuscripts.[4]

Moreover, the Brahatsamhita,[5] a 6th century Sanskrit encyclopedia, references in Sanskrit literature a large number of cosmetics and perfumes that served mainly for purposes of worship, medicine and sensual enjoyment. *Gandhayukti*, a text dating back to the ancient period, offers recipes for making scents. It gives a list of eight aromatic ingredients, one of them being the rose. The *Gandhayukti* also provided instructions for preparing perfumes, bath powders, incense and talcum powder made of roses.

Rose memoir from my childhood

Being raised in the Himalayan Mountains of India amidst pythons, boas, langurs and tigers, I found the flora as impressive as the fauna. Wild apples and pears, apricots, berries and abundant species of flowers and shrubs adorned the terrain. The region includes the famous forests of Kumaon, known for man-eating tigers, which Jim Corbett could not resist writing about in his classic book, *The Man Eaters of the Kumaon*.[6]

My father Dr. V.R. Rajagopalan was a well-known bacteriologist and veterinarian instrumental in developing the serum for Hoof and Mouth Disease and Rinderpest, diseases of the cow.[7] He was also an avid gardener and grew every vegetable, fruit, herb and flower imaginable on the beautifully terraced layout of the land. With great pride, he would walk with us through the garden and tend to the plants with the love and care they would have received if his own. The flower garden was our pride, too. There were many flowers, but the rose section had an amazing variety – some fifteen species to choose from, our favorites being the musk and damask rose. The fragrance of sweet peas and roses mesmerized our senses. As children, when we played house, I remember eating a meal of rose petals. They had been cooked in butter outdoors on the log fire.

However, the langurs provided a serious challenge, the monkey species coming in herds to attack the garden. At times they would destroy the roses; even the monkeys could not resist the taste and the rose hips that provided their nutrition.

But this life was soon to become history as my father was offered the envied post of Director of Animal Husbandry with the Government of Southern India. So we moved to Madras, now called Chennai, located on the South-East coast.

My ancestral heritage lies in this Tamil-speaking region of Southern India; I belong to a liberal Brahmin family known as the Tamil Iyers or Tamil Brahmins. Brahmins could be aptly described as enlightened people who through Vedic education, good hygiene, spirituality upholding values of dharma, eating healthy vegetarian foods, and practicing yoga attained the zenith in the social strata. According to Sanatana Dharma, the ancient name for Hinduism[8] which stands for eternal values of life and a code of ethics that is mentioned in the Vedic scriptures including Bhagavad Gita[9] and the Upanishads,[10] a person might qualify as a Brahmin by birth, by profession, or by conduct based on ethics and morals, intelligent living, right thinking and pure mind. Therefore, all people including a blue collar worker from America, an African from Africa, or a Chinese citizen from Guanghzhou could become a Brahmin by virtue of following Sanatana Dharma. That is, anyone in the social structure with the Dharmic attribute and behavior could be considered a Brahmin. On the same note, a Brahmin by birth who has given up the attributes or principles of dharma is no longer a Brahmin.

The Tamil obsession with roses

Madras, located in the tropical zone, was hot and sulky yet it was enjoyable to treat yourself to the native fruits of the region, quite different from those of the Himalayan mountains. Most of all, it was impossible to ignore the beauty and intoxicating effect of the South's many colorful and fragrant flowers, most notably the rose.

The family tradition mentioned above was carried on. That is, we started a garden in the South with the little space we had for gardening. The flora and fauna were specific to the region. We would stroll on the street to spot the various trees and plants with magical medicinal properties that we might appropriate. And what abundance! Everywhere you turned, you found neem trees, mango, tamarind and gooseberry trees. The coconut trees, standing tall and firm reaching up to the sky, decked nearly every home. Frankly, I was surprised to find in an urban environment such varieties of plants and trees. The homes took pride in growing guavas, pomegranates, coconuts, mangoes and papayas alongside flowering shrubs such as the rose, jasmine, bullet wood and hexapetalus, also known as champagam or Manoranjitham. Because most flowers exude fragrances in the evening, it was always a delight to take a walk at day's end, for as you strolled along you experienced blissful aromas from fragrant blossoms. We children would play games to identify the fragrance. This way we also learned the names of these plants and flowers.

Soon it became a way of life, living in the region marked by Tamil heritage, to learn the nuances of Brahmin Tamil culture. For instance, we discovered how each of the plants, flowers, herbs and animals was intertwined with spirituality. This was evident even in personal adornment. The women of the South, young and old, took pride in decorating their long braided hair with flowers. The most sought after were the jasmine and the rose. Different varieties of jasmine have their own unique fragrances. But among the roses the Panneer Rose was the most desired. Panneer roses are plucked during the day while still in bud form and strung into swags or garlands. The thread used for the rose garlands and flower swags is a natural string that is taken out of a banana stem. The outer dried layer of the stump peels and the strings can be pulled out of it. It is an art where only the fingers do the trick as no needle is used in the process. By evening the buds open up and delight people with their fragrances.

The obsession for roses is beautifully rendered in this poem:

ஒற்றரைரோஜா

தரை தொடாத
விழுதுகளில் ஊஞ்சலாடுகிறது...
அவள் கூந்தலில்
ஒற்றரைரோஜா

Single Rose

Oh single rose
That which never touches the floor
swaying sublimely on the locks
On her braid
A single rose.

Rose markets

It is an awe-inspiring experience to stroll through the amazing wholesale vegetable and flower marketplace called Koyambedu in Chennai. Practically every fruit, vegetable and flower of the region can be found here. Acres of dealers and flocks of bargainers start well in the wee hours of the morning haggling and bargaining like a scene on the stock market floor. It is not rare to see shoppers pushed unwillingly in all directions, but, steered correctly, you can witness the abundance of so many amazing and unusual plant gems. They adorn the path that leads to this magnificent rose flower market guaranteed to steal your heart. The fragrance tantalizes every breath to ecstasy. No wonder it is known as the flower of the heart! Of manifold varieties, one section has the fragrant kind intended for temples. The vendors, however, possessive of their divine flowers, allow no one to pick or smell them. After all, an offering to god must be pure from the heart and should not be given with distrust. You give to God with a euphoric mind set; any offering is like a gift to friends or even ideally to strangers, bestowed with compassion and love.

The flower markets of Southern India are a-buzz with the hullaballoo of evening happenings. Most markets surround a temple. Their aroma permeates the air as people hustle through to buy flowers – for hair, for the deities, for rituals, for weddings or to honor a special someone with a garland of roses. Perfume shops close by marvel with the essence of these fragrant flowers and plants. Along with florists you find fresh vegetable shops, pan or betel leaves, and herbs and spice shops. What a scene! Ladies in colorful saris bargain with vendors. Customers wear roses, jasmine or mixed flower swags, and smiles are so jubilant after the deal. Before heading home, shoppers make the final stop a visit to the temple deity.

The Rose and Divinity

Hindus do not practice idol worship nor do they believe in many gods. Rather, the attributes of human nature and the elements that manifest in humans are represented in the form of deities, and the deities represent one supreme reality, the power of the source.

The deities in temples symbolize the divine attribute of the Brahman (god head). The divine concept is personified as wealth and prosperity in the Goddess Laxmi; Shakti or power is the Goddess Parvati; courage takes the form of Anjaneya, the Monkey God; Ganesha, the obstacle remover, is the Elephant God; and temples exemplify the elements of nature, water, fire, earth, sky or ether and air. God Shiva represents the five elements just as life exists due to their five manifestations. In Sanatana Dharma, God resides within you, so in essence when you visit these temples, you are reiterating to yourself the importance of self-development in reaching the higher goal of eternal bliss through "dhyanam" or meditation.

On such occasions, offering precious, fresh aromatic flower is the norm. The rose or roja in Tamil means the rose color, pink and fragrant, that corresponds to the legendary panneer rose of Southern India. Woven into rose garlands to adorn the temple, the sweet panneer rose must be of the highest quality because only the top of the line and most aromatic may be offered to the gods, and flower offerings are the most sacred of all. In ancient tradition, each temple had its own flower garden known as the "Nandavanam." A priest assigned to horticultural service alone would bathe and then, wearing a wet loin cloth, pluck the scented blooms. Any that fell to the ground were discarded. The flowers would then be strung into garlands by devotees whose lives had been dedicated to serving the higher power, the universal energy source. Thereafter priests would grace the deity with rose garlands and other flower and herb chains, for instance, strings of red roses to adorn and glorify Goddess Parvati whose attribute is Shakti (power) and who is the consort of the universal energy that is Shiva. The perfume opens our senses to subtlety, thereby enhancing the power of the third eye also known as Angya chakra. The womenfolk enhance the power of the third eye with the kumkumam – the well-known red dot on their fore-heads –, its vermillion made with an elixir of turmeric herb.

The Rose and its cosmic connection

Since ancient times in Southern India, the chakra meditation has been the key to good health and longevity. Human beings exist in this universe deriving energy from two sources: the cosmos and mother earth. The physical body is enveloped by an auric field in which seven major chakras, invisible to the naked eye, are present. Each chakra is associated with an endocrine gland and controls specific organs. Disease is the result of the imbalance of chakras, and cure is restoring the balance. In the human body all chakras start to vibrate when the chakra meditation is performed. Residing dormant or asleep at the root chakra is the kundalini, meaning 'she who is coiled serpent power' or the primordial cosmic energy also known as the coiled serpent. The root chakra has invisible roots that ground us in the earth. The body receives earth's energy through the root chakra.[11]

Once Kundalini is awakened, the energy uncoils and starts moving upward through the back chakras of the spinal cord towards the crown chakra where Lord Shiva, also known as the creator, resides. For smooth passage of the Shakti power, i.e. the energy, each chakra is animated by chanting a mantra specific to the chakra. Both the back and the front chakras of the spinal cord are visualized, while breath is taken in through the back chakras and let out through the front chakras forming a garland of roses. The fragrance of the rose resonates throughout the body, healing and energizing. The Shakti ascending reaches the rose energy of the cosmos where the creator Shiva and Shiva-Shakti unite, forming the most powerful mantra known as Ohm. The combined Shiva-Shakti circulates through the chakras. Thus from mother earth and the cosmos, the rose garland makes the body receive the energy believed to cure all diseases and lead to higher awareness.

The mystical red rose is perceived as the all-encompassing cosmic concept of existing matter and space - where the sound of the cosmos resonates through the drums of Shiva as he performs the cosmic dance[12] of annihilation, removing evil and worldly sorrows, releasing souls from the bondage of illusion and preparing for the new creation. Every thorn in the rose represents worldly sorrows, while the blooming rose suggests the blossoming of a new creation out of the sorrows.

The Rose in Tamil weddings

It is therefore not surprising that the rose plays an important role in Tamil weddings. The groom and bride are given the status of Lord Vishnu, the sustainer, and Goddess Lakshmi, the holder of wealth and prosperity, and the unity of their two souls is assured when they are adorned with rose garlands during the ceremony. The groom is always welcomed with a garland of roses.

The famous Indian epic Ramayana[13] describes the lovely ancient wedding ceremony of Sita. The day comes when Sita, with skin soft as the petals of a rose, of beauty unsurpassed, dressed in silks, gold and gems, her locks decked with roses and fragrant flowers, waits with a garland of roses for a worthy suitor who must prove his skills. Her many suitors, awestricken by her splendour, attempt to break the celestial bow with powers unimaginable.

Finally, a prince named Rama steps in, with aura spreading through the seven heavens, for he is descended from the "Raghu-vamsa" (the sun); and his illuminated visage, grace and courage suggest that this union had been sealed in paradise. With one finger Rama picks up the bow and breaks it in two. All attendees, including the rishis who are the enlightened people, kings and sages, rejoice, and in a ceremony called "Swayamvaram" (marriage) Sita garlands Rama as a gesture of acceptance. In Sanathana Dharma, during ancient periods, simply exchanging garlands solemnized a wedding.

Weddings at present are quite elaborate. Preparations include banana tree stumps with bananas and banana flower intact tied at the entrance of the nuptial arena to signify everlasting bounty and healthy progeny. Furthermore, never-fading mango and screw pine leaves enwrap the entrance, just as kolam art[14] welcomes prosperity and splendid roses serve as offerings and benedictions.

Rose petals and rice mixed with turmeric are given to all guests at the end of the ceremony. It is said that this sacred event is witnessed by Devas (godheads residing in the heavens). As the newly-weds prostrate with respect to all the elders who have come, who in essence are believed to be the Rishis and Devas, guests bless the couple by strewing the distributed rose petals and smearing the rice with turmeric powder on their heads.

Rose fragrance brings joy to the wedding arena. Striking in embroidered jari silks, flowers in their hair, adorned with golden jewelry and gems, young teenage girls and their million dollar smiles welcome guests by sprinkling rose water from intricately carved silver vessels. To calm and soothe the senses they also use sandalwood paste and turmeric in enchanting silver bowls while palm sugar candy heightens the sweetness of the day, and flowers like roses and jasmine are showered in abundance on all celebrants.

Rose water

Rose water, moreover, is an appropriate offering to the deities. The gods are sprinkled with it after a ritual known as "abhishekam" during which their likenesses are bathed in milk, honey, yogurt, ghee (clarified butter) and fruits. The divinities are then bejeweled with ornaments and roses while reverent worshippers chant Vedic mantras, that is, sacred hymns in praise of the attributes the deity stands for.

Rose water is made with great care, unadulterated and pure for the offerings. It can be produced at home with any variety of fragrant roses. Clean petals without stems and leaves are soaked in pure distilled water, and hours later, they turn translucent. The water absorbs the color and fragrance. This liquid is then stored in airtight glass containers although custom also recommends vessels of silver. Liquid scent that is not exposed to light retains its fragrance longer. This is indeed a very old tradition: the earliest distillation of attar –

or essence water--, was mentioned in the Ayurvedic text Charaka Samhita.[15]

Not exclusively reserved for the divine, rose water is also used as a facial and body splash; it is healing, rejuvenating and calming. Hair, when washed and then rinsed with rose water, is left bouncy and fragrant. Moreover, rose water has shown success as a remedy for Attention Deficit Disorder and Attention Deficit Hyperactivity Disorder, and this in turn fits well with Brahmin Tamil culture as a way of life comprised of yoga, good hygiene, aroma therapy, and food that was also spiritual and medicinal. Independent of religion, social class or politics, this chosen lifestyle promoted a healthy body, mind and soul. And rose aroma therapy, in particular, maintained youthfulness and calm. In sum, the culture stressed precaution to avoid disease because 'prevention is better than cure'.

Kingdoms and Roses

I wish you the good fortune to visit Madurai in Southern India, an auspicious city dating back over 2500 years and the home of Meenakshi Amman Temple. Goddess Meenakshi is an avatar of Parvati, the consort of Shiva. Although the original edifice was destroyed during the Muslim invasion of India somewhere around the 10th century, it was rebuilt in the 12th century. Besides the grandeur of the temple, what is most impressive is the importance the town accords to flowers, especially the rose and the lotus. Flower markets near the shrine are crowded and flurry with activity in the evenings where, for instance, when fed and given money, the temple elephants reward devotees with rose garlands.

But these are not just any garlands. This region is home to rose attar or pure essence of rose. Storekeepers take pride in showing off the scent. By reaching out to help and guide the customer, these pleasant merchants exude their own pleasing aura that inevitably coaxes you into a few purchases of their precious fluid.

Hundreds of years earlier, however, you would have experienced a similar phenomenon. From the 14th to the 16th centuries, one of the greatest flourishing dynasties was the Vijayanagaram Kingdom of Southern India. During this epoch, cultivation of roses was encouraged, and growers were rewarded with lower tax assessments. Aristocratic men and women adorned themselves with roses and sprayed themselves with rose perfumes. Two Portuguese travelers,

Domingo Paes and Fernao Nuniz[15] who visited this kingdom around 1520 CE were awestruck by its architectural beauty and wrote about it in their chronicles, including their admiration for the vast number of roses. They saw the king, Sri Krishna Deva Raya (1509-1529 CE) dressed in white clothes embroidered all over with golden roses, and the horses, too, sported roses and other flora on their heads and necks.

Furthermore, the rose was not confined to fabric. At the top of the cross-timbered pillars in the king's carved chamber, the explorers found roses and lotus, all beautifully executed in ivory and more rich and beautiful than anywhere else. The Portuguese also saw plantations of roses in the city of Bisnaga and throughout the countryside. The roadside bazaars of Hampi, the capital, had many shops whose baskets were laden with the blooms.

The author of the chronicles of Nuniz writes that Hindus revered as a Saint the King of Delhi who had advanced towards the Empire of Vijayanagar. They say that once when he was offering prayers to God, roses from heaven cascaded down on him.

Perhaps this inspired St. Therese of Lisieux, known as "the little flower of Jesus," who promised that she would bless the world by letting "fall from Heaven ... a shower of roses."

In conclusion, the Tamils of South India eat roses, breathe roses, pray with roses, heal with roses and just live happily with roses, basking in their unique fragrance, following Sanathana Dharma, which comprises the Eternal values of life, allowing roses to fall from heaven, to bless them, to rejuvenate them and to reach the higher consciousness that is the Rose.

References

[1] The ancient Ayurvedic text written by Charaka is known as Charaka Samhita. His scholarly work, a compilation of several writers, guides practice in modern day Ayurvedic treatments.

[2] Abdur-Razzaq was the ambassador of the ruler of Persia to Calicut, India, from 1442 to 1445. During this time he chronicled his journey in Matla-us Sadain wa Majma-ul-bahrain or *The Rise of Two Auspicious Constellations and the Confluence of Two Oceans*. It described the life and events in Calicut under the Zamorin and also the ancient city of Vijayanagar in South India, at Hampi.
Beller-Hann, Ildiko. *A History of Cathay: a translation and linguistic analysis of*

a fifteenth century Turkic manuscript. Bloomington: Indiana University, Research Institute for Inner Asian Studies. 1995. 11.

[3] See Rig Veda, 107; healing plants: RV X-97.

[4] K. Gode. *Studies in Indian Cultural History.* Vol. 1. NP: np, 1961.

[5] „Preparation and testing of perfume as described in Brahatsamhita." Sachin A. Mandavgane, Holey and J. Y. Deopujari. *Indian Journal of Traditional Knowledge.* Vol. 8 (2) April 2009. 275-277.

[6] Jim Corbett. *Man Eaters of Kumaon.* London: Oxford UP, 1944, 1952, 1959.

[7] V. R. Rajagopalan. *Pasteurellosis in white mice and diseases of the cow.* Indian Science Congress 1939, 1949.

[8] Jagannath Dixit, Jay Dixit. *Chitpavanism: A Tribute to Kokanastha Brahmin Culture.* NP: Dixit, 2004.

[9] Ekta Singh. *Caste System in India. A Historical perspective.* NP: Gyan books, 2005.

[10] Chinmaya Mission West. "Hinduism that is Sanatana Dharma." publications@chinmaya.org. Email to the author. 2007.

[11] Sanatana Dharma refers to activities that pertain to the soul, not to the body. Spiritual culture is recommended for anyone and everyone, whether in the body of a man or a woman, young or old, married or unmarried. In any situation, the natural proclivity of the soul is to serve the super soul, the Supreme Personality of Godhead. "It is not possible for the living entity to be happy without rendering transcendental loving service unto the Supreme Lord." A.C. Bhaktivedant, a Swami Prabhupada. *Introduction to* Bhagavad-gita *As It Is.* NP: NP, 1989.

[12] *Upanishads* are a sacred collection of writings from India's oral history that have been passed down. It sets forth the prime Vedic doctrines of self-realization, yoga and meditation. The *Upanishads* are summits of thought on mankind and the universe, designed to push human ideas to their very limit and beyond. They give both spiritual vision and philosophical argument to attain self-realization.

[13] Harish Johari. *Chakras: Energy Centers of Transformation.* NP: Destiny Books, 2000.

[14] Vasant V. Merchant, Siva-Nataraja. "The Lord of the Dance, Drama and Music; Siva – the cosmic Dancer, Transformer, Liberator." *International Journal of Humanities and Peace.* Available from amazon.com

[15] K. M. K. Murthy, trans. *Valmiki Ramayana, The First Epic Poem of India.* NP: NP, 2008.

[16] Kolam is drawn in front of the home every morning before the sun rises. Kolam art consists of ancient symbols sketched to energize the house. An inspiration for sacred geometry and Reiki symbols, this art form empowers and energizes the women folk who alone are bestowed the right to draw the symbols of kolam art. It is produced daily and on auspicious occasions such as weddings and festivals. Rice flour is used either in a dry form or as a thin paste to craft the geometric shapes and loops around a pattern of dots, the artists thereby exercising their dexterity, coordination and math skills. The area is washed with cow dung water to sterilize the entrance before the art is created. Kolam is then drawn to welcome wealth and prosperity into the house. Kolam is also placed in front of the family alter, on the steps and at the entrance door.

[17] Michael Dick. MS. *Ancient Ayurvedic Writings.* Albuquerque, NM: The Ayurvedic Institute, 1998. Charaka Samhita is believed to have arisen around 400-200

BCE, felt to be one of the oldest and most important ancient authoritative writings on Ayurveda. See also Dastur, *J. F. Everybody's Guide to Ayurvedic Medicine.* Bombay: D. B. Taraporevala Sons & Co., 1978.

[18] Robert Sewell, Fernao Nuniz, Domingo Paes. *A Forgotten Empire: Vijayanagar. A contribution to the history of India* (includes a translation of Domingo Paes and Fernao Nuniz. *Chronica dos reis de Bisnaga.* 1520 and 1535). NP: Adamant Media Corporation, 1982.

9

Personal Reflections on the Rose in Vietnam

Vickie TườngVi Eaton

A tropical country, Vietnam is blessed to find roses growing almost everywhere. Symbolically, the rose represents love, passion and grace. It has many varieties, colors and meanings. Traditionally, red roses provide the hue of choice for weddings and celebrations although, in recent years, in arrangements for all occasions, bouquets in mixed colors have become fashionable.

In addition to decorating with whole flowers, people enjoy making home-made lotions of rose petals, even in the absence of official sanction for the many domestic formulas. This suggests a need to increase research on roses and their application in Eastern medicine, but efforts already undertaken in Viet Nam to honor this important flower are worth surveying. In his book *Những Câythuốcvà Vịthuốc Việt Nam*, Professor Đỗ Tất Lợi[1] refers to the medicinal aspects of the rose family. He describes harvesting, for instance, Rose Multiflora, Rose Canina, and Rose Lavigata and their branches which, after drying, are used to treat conditions such as skin eczemas and irritations or diarrhea. The elixirs also have a preventative function as a dietary supplement helping sustain the kidney, liver and blood.

Now, medicine isn't the only association the flower evokes in Vietnam. The Vietnamese also think a lot about roses' various hues and, not unlike the West (just ask your florist), they attach meanings to the number of blooms in a single bouquet. But most significant today, when Vietnamese consider the rose, we associate it first with an image of our mothers and the blessed life we have been privileged to lead with them.

This connotation is relatively new, however, inspired by a specific encounter between the Zen Master **Thích Nhất Hạnh** and an association between red and white carnations on Mother's Day in Japan. He brought the idea of the flower and the mother – in Vietnam, the "pink" and the "rose" – back to our country, and published a tender, emotional article as well as a poem on Le Vu Lan – Parents' Appreciation Day – in 1962. In this tribute to his deceased mother, he offered a gift to people whose parent still lived, reminding them of their good fortune and urging them to treasure their mother while she remains alive on earth.

Thích Nhất Hạnh's publication inspired song writer **Phạm Thế Mỹ** to compose "Rose Pin" – meaning a broach – to celebrate motherhood. According to Buddha's teaching, we must honor our father and mother, obey and respect them always, and among the many songs, poems, folk melodies and tunes in praise of parents, "Rose Pin" is appreciated, even today, as one of the best lyrics in Vietnamese music.

This popularity may in part be due to the double meanings of the word *rose* and hence "Rose Pin" in Vietnamese: the floral rose is **hồng**, but the color pink is also **hồng**. Now, in Japan, the flowers that had inspired Thich Nhat Hanh were a white carnation if the mother is deceased and a red one if she still lives. (You'll find the story below). Thus, I suspect that both **Thích Nhất Hạnh** and **Phạm Thế Mỹ** were happy to retain the double idea of a 'rose rose' – that is, the pink rose – to capture the symbol of color and life while retaining the symbol of white as death, thereby showing the depth of our duality, our vitality and mortality. After all, red and white make pink; and while the pastel retains the white, its color transcends it.

Now, here's the story. Poet, writer, and teacher, **Thích Nhất Hạnh**, one of the most renowned Zen Masters alive today, was born in 1926 in **Quảng Ngãi**, Central Vietnam, and entered **Từ Hiếu** Temple at age 16. Ordained a Buddhist monk in 1949,[2] he travelled all over the world, learning and teaching about peace and meditation. In Medford, USA, in 1962, he wrote about losing his mother:

> In the West, we celebrate Mother's Day in May. I am from the countryside of Vietnam and had never heard of this tradition. One day, I was visiting the Ginza district of Tokyo with the monk Thien An, and we were met outside a bookstore by several Japanese students who were friends of his. One discretely asked him a question, and then took a white carnation from her bag and pinned it on my robe. I was surprised and a little embarrassed. I had no idea what this gesture meant and I

didn't dare ask. I tried to act natural, thinking this must be some local custom.

When they were finished talking (I don't speak Japanese), Thien An and I went into the bookstore and he told me that today was what is called Mother's Day. In Japan, if your mother is still alive, you wear a red flower on your pocket or your lapel, proud that you still have your mother. If she is no longer alive, you wear a white flower. I looked at the white flower on my robe and suddenly I felt so unhappy.

I was as much an orphan as any other unhappy orphan; we orphans could no longer proudly wear red flowers in our buttonholes. "Those who wear white flowers suffer, and their thoughts cannot avoid returning to their mothers. They cannot forget that she is no longer there. Those who wear red flowers are so happy, knowing their mothers are still alive. They can try to please her before she is gone and it is too late." I find this a beautiful custom. I propose that we do the same thing in Vietnam and in the West as well.[3]

His poem translated into English[4] is quoted below. In 2005, the article and the poem were published in Thich Nhat Hanh's book *A Rose for Your Pocket, an appreciation of motherhood.*

When I was a child I heard a simple poem about losing your mother, and it is still very important for me. If your mother is still alive, you may feel tenderness for her each time you read this, fearing this distant yet inevitable event.

That year, although I was still very young
my mother left me,
and I realised that I was an orphan,

everyone around me was crying,
I suffered in silence...
Allowing the tears to flow,
I felt my pain soften.

Evening enveloped
Mother's tomb,
the pagoda bell rang sweetly.
I realised that to lose your mother
is to lose the whole universe.

We swim in a world of tender love for many years, and … are quite happy there. Only after it is too late do we become aware of it.

   That Thich Nhat Hanh should have inspired songwriter **Phạm Thế Mỹ** (1930-2009)[6] is not surprising. Born in northern Vietnam, Mỹ migrated to what is now Ho Chi Minh City and studied music at the Saigon National Academy of Music and Drama. Having read the monk's article and poem, he wrote "Rose Pin" while in jail for his involvement in fighting for the Buddhist movement (1965-66).[7] The beautiful yet natural lyric associates a mother's love with simple, everyday country foods like bananas, sweet rice and sugar cane. The immortal ballad also confirms that the pink rose pinned on a lapel expresses the emotion, love, admiration and honor that motherhood deserves while a mother still lives, because her life is the greatest source of joy. Here is a partial translation.

   Một bông Hồng cho em (A rose for you)
   Một bông Hồng cho anh (A rose for me)
   Và một bông Hồng cho những ai (And a rose for those who)
   Cho những ai đang cònMẹ
              (For those who still have a mother)
   Đang còn Mẹ để lòng vui sướng hơn
              (It is so much joy to have Mother with me)
   Rủi mai này Mẹ hiền có mất đi
              (If somehow Mother leaves me)
   Như đóa hoak hông mặt trời
              (It would be like a flower without the sun)
   Như trẻ hơk hông nụ cười
              (It would be like a child without a smile)…
   …
   Mẹ, Mẹ là doing suối dịu hiền
              (Mother, you are a gentle stream)

Mẹ, Mẹ là bài hát thần tiên (Mother, you are the fairy song)

Mẹ, Mẹ là lọn mía ngọt ngào
>   (Mother, you are the sweet sugar cane)
Mẹ, Mẹ là nải chuối buồng cau
>   (Mother, you are the banana, the betel nuts)
Là tiếng dế đêm thâu
>   (You are the cricket sound in the deep night)
Là nắng ấm nương dâu
>   (You are the warm rays on the field)
Là vốn liếng yêut hương cho cuộc đời
>   (You are all the source of love in life)

óa hoa màu hồng vừa cài lên áo đó anh
>   (A rose pin for your dress for you)
óa hoa màu hồng vừa cài lên áo đó em
>   (A rose pin for your dress for me)
Thì xin anh, thì xi nem   (I am so happy to have Mother)
Hãy cùng tôi vui sướng đi.
>   (I celebrate Mother, I honor Mother)[8]

Viet Nam Roses in Commerce

At 1,500 meters, Langbiang Mountain and its town of Da Lat,[9] nick-named Vietnam's "Eternal Spring City," offer cool, moderate high-land temperatures to encourage the many flowers, plants and trees that thrive there. Gorgeous roses including cultivars grown on the plateau exude an especially long-lasting fragrance.

Moreover, from start to finish, in response to the blossom's high economic value, Da Lat's growers take special care in rose pro-duction. After harvesting, they transport the blooms to Ho Chi Minh City (formerly Saigon) for sale and distribution.[10]

Now, because roses represent romantic love and passion, more innocent flora is preferred for Buddhist altars – fresh white tube-roses or gladioli, for instance – to honor Quan Jin and the ancestors, as Buddhists' worship requires purity and respect for Buddha.

Christians, however, for many historical reasons favor roses. Popular in Vietnamese churches, the flower represents Christ's resurrection on the 3rd day. Seven days a week in front of Saigon

Cathedral, worshippers find innumerable vendors hawking fragrant bouquets.

Homes, too, as small as they are, include boxes in which to plant roses, among them the very well-liked trellis rose **Tu'ò'ng vi** – my name, by the way, Vickie Tuong Vi Eaton. In the family of Lagerstroemia (Lagerstroemia speciosa), this easily-cultivated vine plant is especially decorative as it climbs walls, adapts to tiny spaces and shares an intoxicating fragrance. My understanding is it migrated from the North to the South in 1948 when Vietnam was cut in half, divided between the two opposing political parties.

This division and the ugly period in our nation's history that followed were not without negative consequences for the rose. Yes, roses remained attractive, but in war-torn Vietnam the flower lent its name to the other side of beauty, too, namely death. You have heard of American soldiers and Vietnamese "bar girls"? In the 1960s, rumors spread of a deadly STD called Vietnam Rose or the "Black Pox."

Varying descriptions of the rose disease emerged in different regions. Called "Black Syphilis" in Thailand, it was also known in Northern Iraq in the first Gulf War, and although the tales have not been verified, they likely include some moments of truth. Said to have no treatment or cure, the illness caused its victims, mainly men, to lose their penises; they were said to rot and fall off. It was further rumored that infected soldiers were extradited to an island where they would spend the rest of their lives without their family's ever learning of their fate.

So what caused this plague? Possibly some resistant strain of gonorrhea manifesting as genital warts or other lesions, though diagnoses fail to correlate with any medically recognized term.

As to the fate of the "bar girls" thought to initiate the Vietnamese Rose, those infected were never discussed. During the war, many in the countryside, forced to leave their homes, found themselves in Saigon city with no means of support. Impoverished teens often felt compelled to sell their bodies to survive and to support families left behind in villages.

Whether or not the details concerning the Vietnam Rose are true, tales about young Vietnamese women and lonely American soldiers who found comfort for a night in bars are all too plentiful. A few are beautiful love stories, but most are tragedies with mixed children born, abandoned and abused by disoriented families, the soldier fathers and partners killed, and the women, too, abandoned or dead. The Broadway play *Miss Saigon* reflects one such drama after the war.

Rose Petals and Homemade Rose Water[11]

Ironically, far from erasing the beauty of the rose in Vietnam, the war increased opportunities to deploy the petals in first aid and beauty remedies.

How? Here's the formula you have been waiting for. To produce rose water, place fresh petals from five or six roses in three to four cups of fluid, bring the liquid to a boil, reduce the heat and then simmer for twenty minutes. Allow the juice to cool before refrigerating it in dark glass bottles. The petals remain in the brew. It can then be applied as a lotion to tighten pores, to clean the skin before applying face cream, or to help prevent sunburn. The elixir can also reduce acne. Your coiffure, too, can benefit if you soak it in rose water twenty minutes before shampooing. The essence, by bringing out luster and shine, is especially suitable for damaged hair. Here's a popular recipe for rose shampoo:

Take 1/4 cup of the rose water you have made and mix it with 1/4 cup of red wine, three eggs and a cup of white vinegar. To restore the hair's silkiness, use this mixture instead of shampoo.

Men, by the way, use rose water as an aftershave to ease skin irritation, and in case you need to lighten the dark circles under your eyes, rose water mixed with cucumber will do the trick. This multipurpose brew has also been reported useful against menstrual cramps, bad breath, sore throat and children's coughs.

In sum, beauty is as beauty does. In Eastern medicine, roses do good, inside and out.

Thanks to Tobe Levin for editing this chapter.

References
[1] See http://www.ebook.edu.vn/?page=1.27&view=8770 e-Book. p 892-896. Retrieved 14 December 2014.
See also http://www.vinabook.com/nhung-cay-thuoc-va-vi-thuoc-viet-nam-11i11980.html Accessed 14 December 2014.

[2] "Thich Nhat Hanh." Wikipedia. http://en.wikipedia.org/wiki/Thich_Nhat_Hanh. Retrieved 28 September 2014.

[3] Thich Nhat Hanh. "A Rose for Your Pocket." http://www.buddhismtoday.com/english/others/007-tnh-rose%20for%20you.htm. Retrieved 28 September 2014.

[4] Here in the original and English translation. See Thich Nhat Hanh. *Để dâng mẹ và để làm quà Vu Lan cho những ngườ inào có diễm phúc còn mẹ.* Medford: HoaKỳ, tháng-tám, 1962. http://www.budsas.org/uni/u-vbud/vbpha087.htm. Retrieved 28 September 2014.

[5] "A Rose for Your Pocket." http://www.buddhismtoday.com/english/others/007-tnh-rose%20for%20you.htm. Retrieved 28 September 2014.

[6] "Phạm Thế Mỹ." http://vi.wikipedia.org/wiki/Ph%E1%BA%A1m_Th%E1%BA%BF_M%E1%BB%B9
See also http://giaitri.vnexpress.net/tin-tuc/gioi-sao/trong-nuoc/tac-gia-bong-hong-cai-ao-qua-doi-1901772.html. Retrieved 28 September 2014.

[7] "Bông hồng cài áo." http://vi.wikipedia.org/wiki/B%C3%B4ng_h%E1%BB%93ng_c%C3%A0i_%C3%A1o. Retrieved 28 September 2014.

[8] "Bông hồng cài áo." http://www.youtube.com/watch?v=Q7n3LOSJhSY. Retrieved 28 September 2014.

[9] Before 1975 the capital of the French colony of Cochin-china and the largest city in Vietnam, Saigon in the 1960s was called "The Pearl of The Far East." After April 30, 1975, Saigon merged with Gia Dinh Province and changed its name to Ho Chi Minh after the communist party leader who won the war in 1975. Viet Nam then became a communist country. The name Saigon is still very popular, however, not only because it had been in use for over 300 years but also since it was so well known in Asia in the 1960s.

[10] "Ho Chi Minh City." Wikipedia. https://en.wikipedia.org/wiki/Ho_Chi_Minh_City Retrieved 14 December 2014.

11 "Rose Cures Bad Breath." In Sử dụng thuốc. http://www.microsofttranslator.com/bv.aspx?from=&to=en&a=http%3A%2F%2Fthuoc-dongduoc.vn%2Ftin-tuc-su-kien%2Fsu-dung-thuoc%2F1759-hoa-hong-chua-hoi-mieng.html. Retrieved 29 September 2014. See also Sử dụng thuốc. "Hồng Hoa Và Hoa Hồng." http://thuocdongduoc.vn/tin-tuc-su-kien/su-dung-thuoc/1492-hong-hoa-va-hoa-hong.html. Retrieved 29 September 2014.

10

The Chinese Rose:
The "Queen of Flowers" in Cathay[1]

Yaping Qian

In the Northern Song Dynasty[2] of ancient China, there appeared a famous official called Bao Zheng (999 - 1062). Commonly known as Bao Gong or Lord Bao, he was intolerant of any corruption, bribery, or dereliction of duty, and took an uncompromising stance against social injustice, thus earning for himself a sustained title – Justice Bao, the most renowned judge of the period. While celebrating his 60th birthday, Bao Gong told his son Bao Gui not to accept any presents. Keeping the father's order in mind, the son first declined Emperor Renzong's gift, then the one from his father's bosom friend, Zhang Kui. Nonetheless, when a commoner representing the ordinary people came to congratulate with a pot of the Chinese rose in his arms, Bao Zheng took the flower with alacrity. Why did he refuse the nobles' luxurious tributes but favored the Chinese rose? The underlying messages are not hard to discern: symbolic of the public praise and veneration for the integral Justice Bao, the Chinese rose was, to him, a mirror, an inspiration and the exhortation to cherish an everlasting loyal heart as red as the rose petals to serve the people.

> The early 20th-century plant hunter Ernest H. Wilson dubbed China the "Mother of Gardens" for good reason: The country is home to some 31,000 native plant species, a third more than the U.S. and Canada combined, and plant hunters have avidly collected Chinese species to transplant in North America and Europe. Gardens throughout the world today showcase flowering plants – rhododendrons, forsythias, magnolias, camellias, primroses, viburnums, and many others – that originated in China.[3] (Susan K. Lewis, 112)[4]

In 1913, E. H. Wilson's *A Naturalist in Western China* came out, in which the notable botanist introduced a large range of about 2,000

Asian plant species to the West. The book was republished sixteen years later titled *China, Mother of Gardens*. Exactly as "Chinese" Wilson[5] uncovered during his eleven years' travel, exploration and observation in remote parts of China, China is unquestionably a flowery kingdom with an unparalleled richness of biodiversity. "HuaHsia" (also "Huaxia" or Cathay) is the ancient name for China, and "Hua," meaning prosperity in ancient times, was equivalent to "hua" (referring to flowers). In effect, the origin and evolution of Cathay[6] have maintained an intimate connection with flowers. Consequently, as a nation planting, appreciating and making use of flowers for more than 3,000 years, the Chinese people have created a colorful and profound floriculture. And in that long history, numerous valuable floral records have been kept in many Chinese classics, for instance, in China's first poetry collection *Shih Ching (The Book of Songs)*, *Qunfangpu[7] (Beautiful and Fragrant Flowers)* in the Ming Dynasty (1368 - 1644) of ancient China, the *Zhongguohuajing[8] (China Flora Encyclopedia)* in the modern age, etc., which provide lively evidence of the natural affinity between the Chinese nation and flowers.

The Three Chinese Sisters in Chinese Floriculture

Rosa chinensis (Yuejihua/Yueji), Rosa rugosa Thunb (Meigui) and Rosa multiflora (Qiangwei) in pinyin, commonly known as multiflora rose, baby rose, many-flowered rose and seven-sisters rose, are triplets of the genus Rosa. Although the three species' genetic proximity makes it hard to differentiate one from another, some slight dissimilarity exists. The perpetual-flowering Yuejihua usually grows single isolated flowers bigger in diameter on top of the branches with sparse prickles. Meigui, in comparison, has only one blooming period, and the stems are generally armed with densely packed straight thorns and wrinkly leaves. Apart from its medicinal uses, the fragrant edible Meigui can be extracted for commercial perfumery. Unlike the former two, Qiangwei with its pinnate leaves is a scrambling shrub climbing over other plants though flowering occurs only in spring, too.

These differences were already noted in ancient China. In *Shuo-wen chieh-tzu (Annotation Explaining and Analysing Words*, often shortened to *Shuo-wen)*, an early 2nd-century Chinese dictionary compiled by Xu Shen (c. 58 c. 147), a Han Dynasty (206 BC 220

AD) scholar, says, "Mei is the best of jade; Gui the prettiest jewel." Thus, it is obvious that Meigui originally bore no relation to flowers in Chinese culture. The term, then, was instead rendered to Rosa rugosa Thunb, although when and why still need further identification. The Ming Dynasty scholar Wang Shimao (1536 - 1588), in his three-volume *Xuepu zashu*[9] *(Horticulture Miscellanies)*, narrates, "[Meigui is] attractive in color, as in flavor ... [is] edible and can be worn as an adornment." The early 17th-century *Qunfangpu* describes Meigui as resembling Qiangwei and having dense prickles, fragrance and color, and it can be drunk with tea, wine and honey, as well. One of the "Four Poetry Masters" of the Southern Song, Yang Wanli[10] (1127 - 1206), opens his poem *Red Rose* (Red Meigui) stating clearly that "Named differently from Yuejihua, she is not linked with Qiangwei either." And Bai Juyi (Po Chu-i) (772 - 846), a celebrated poet of the Tang Dynasty (618 - 907) after Li Bai[11] (Li Bo) (701 - 762), remarked, "However elegant the lotus is, in mud its calyx rises; however attractive Meigui looks, with prickles its stems stretch."

An entry for "Qiangmi," referring to "Qiangwei," first made its appearance in Volume 18 of the pharmaceutical text, *Bencao gangmu (The Compendium of Materia Medica)* published in 1593 during the Ming Dynasty of China. The author Li Shizhen (Li Shih-Chen) (1518 - 1593), one of the greatest Chinese herbalists and acupuncturists in history, clarifies, "For the stems being scrambling and pliable, the wall-climbing herb is thus called Qiangmi." The 17th-century *Huajing,*[12] the earliest and most valuable monographs on ornamental plants in Chinese history, includes a more detailed record of Qiangwei presenting more features of the shrub. Several varieties of Qiangwei are introduced as well in *Qunfangpu.* In one of his poems on Qiangwei, Po Chu-i writes, "Leaning on a tall treillage, opposite the small house, Qiangwei blossoms fully. /When late spring comes, in the center of the courtyard, the flower exhibits its beauty only." Qiangwei's growth habit is here alluded to.

Caohua pu (Botanical Illustration), a collection of poetry and essays by the 16th-century dramatist, writer and encyclopedist of the Ming Dynasty Gao Lian (1573 1620), and Li Shih-Chen's *The Compendium of Materia Medica and Beautiful and Fragrant Flowers* all called the bloom Rosa chinensis Yuejihua, while *Huajing and Zhiwu shiming tukao*[13] *(The Illustrations of Plant Names)* addressed the flower Yueji for short. *The Compendium of Materia*

Medica already classifies it in the genus Rosa and records, "Everywhere Yuejihua is grown and transplanted."

Even today in modern China, on naming the hybrid varieties of roses, a consensus has not yet been reached, and a portion of professionals insists on calling the new types modern rose or Yuejihua for short. Whatever dissensions there are, the triplets cannot wholly replace one another, which is a rock-solid fact. In the West, however, the three species are commonly acknowledged as "rose" without much distinction, which reminds us of the popularly quoted part of a dialogue in Shakespeare's play *Romeo and Juliet*. In reference to Romeo's name, Montague, Juliet says, "What's in a name? that which we call a rose/By any other name would smell as sweet." It is self-evident that names are simply symbols denoting nothing in a serious sense, and therefore, whether to address a flower Yuejihua, Qiangwei or Meigui, in essence, leads to no harsh consequences.

Chinese Birthmarks

In the more than 20,000 cultivars of the genus Rosa family, Yuejihua, Meigui and Qiangwei stand alone. Why? Strong resemblance of the three is one reason. Nevertheless, it is the Chinese birthmarks on them that have tied them closely with each other.

We see this in the 20th-century discovery of "abundant fossils of plants and insects [that] have been found in this [Chinese] coalmine. Wood of Piceoxylon fushunese, Chamaecyparioxylon chinense, Juniperoxylon chinense, etc., was reported from the Lizigou Formation. Over 70 species of leaves and fruits were found in the Jijuntun Formation, including Metasequoia, Keteleeria, Gingko, Cercidiphyllum, Sequoia, and Glyptostrobus."[14] Among the plant leaf fossils, some that date back about 60 million years bear a wonderful likeness to the leaves of modern roses. And then, in Linqu, Shandong of China, in the same century, more such plant leaf fossils dating back 12 million years were discovered, which, together with the fossils evacuated earlier, have been identified by paleontologists as the progenitor of modern roses.

"In 3,000 B.C., in what is now Iraq, the Sumerians created the first written record of the rose. Sappho, in her 600 B.C. 'Ode to the Rose', referred to this beauty as the queen of flowers, a reference still popular today."[15] It has been proved, however, that wild Yuejihua flourished in the northern region of China 3,000,000 years ago.

Although Yuejihua has a mysterious origin, and there is no evidence of how it was first developed, it has been firmly held that it is the product of a rich culture of ingenious people. For instance, according to a Chinese legend, in the Shennong Era[16] (between 5,000 and 10,000 years ago), the primitive Chinese even began to domesticate wild Yuejihua. In the Han and Tang Dynasties, Yuejihua were widely cultivated, and planting Yuejihua was commonly seen in the Ming Dynasty. As early as the 16th century, Yuejihua was prominently introduced into Italy, first through the famous "Silk Road"[17] and then in 1789, via India. Its cultivation then spread to the U.K., the U.S.A. and other countries, where Yuejihua blended with western roses and was appreciated for its beauty, fragrance and long blooming period. The arrival of the Chinese roses changed the rose world fundamentally, broadening the scents of roses and even transforming the form of the flower. Graham Thomas[18] (1909 2003) believes that Yuejihua is the species upon which modern roses are built. Since then, because of this, Yuejihua in Europe has been endowed with an emblematic name— Chinese rose. Hence, doubtlessly, China, in which Yuejihua originated and flourished from time immemorial, is the distribution center and the creditable birthplace of the Chinese rose.

In classical Chinese poetry, best-known works on Qiangwei do not go unnoticed. In the Southern Dynasty, for example, Xie Tiao (464 499), Bao Quan (? 551) and Liang Jianghong (?) all composed poems titled *Ode to Qiangwei*, in which the poet Liang depicts, "Bred in the courtyard, Qiangwei flourishes with exuberant stems and leaves. Unshaken by hands, the scent already diffuses; untouched by wind, the petals fly about by themselves." Xie Tiao says, "Low shrubs are burdened with luxuriant leaves, slight scent oozes out naturally. Ever since the calyx first sprang, early blossoms are arrayed in pale red."

The lines indirectly prove that domestication of wild Qiangwei began 1,500 years ago in ancient China. Meigui as a reference to the flower was previously recorded in *Xijing Zaji (Miscellaneous Records of the Western Capital)*, a collection of short semi-historiographical stories from the Former Han Dynasty (206 BC 9 AD) court, and in which it is illustrated, "Meigui grows in the Leyou Garden, under which alfalfa luxuriates." The Garden here is a royal park built for Emperor Wu of Han, the seventh emperor of the Han Dynasty of China, ruling from 141 BC to 87 BC. The fact that by then Meigui ranked with other rare flowers is therefore certified.

With Chinese blood in their veins, Yuejihua, Meigui and Qiangwei are hence known as the three Chinese sisters in the genus Rosa, this large botanical family. Without a doubt, the triplets have enriched and enhanced the floriculture of China as well as that of the world.

The Chinese Rose in the Rose Culture of China

In time-honored history, a far-reaching rose culture profoundly embedded in traditional Chinese culture has been valued and consecrated, and the Chinese people, from Cathay to modern China, have maintained a close-knit connection to the three Chinese sisters, and particularly to the Chinese rose.

The Chinese Rose & Traditional Chinese Poetry

The ancient Chinese excelled in expressing themselves in the genre of poetry, and in the Tang and Song Dynasties especially, poetry creation, including shi, ci and fu,[19] saw a renaissance. Chinese flowers are a recurring theme in Chinese poetry. As one indispensable organic element of nature, the Chinese rose, with her other two siblings, attracted a variety of poets' attention and admiration.

Chen Xingyi (1090 1138) in the Song era, for example, in his poem *Yuejihua*, chants:

> The rain dampens Yuejihua,
> Spring returns leaning against the rail.
> People from all directions
> The scent more than once inhale.

In a year spring comes only once. However, in contrast, Yuejihua can bloom more than once. In the meantime, the flower's sweet odor has enticed innumerable people from varying locations. Obviously, the poet tactfully writes out Yuejihua's long blooming period.

Another poet, Su Shi (1037 1101), also known as Su Dongpo, one of the major poets of the Song Dynasty, states the same fact in his *To Yueji:*

> Endlessly falling and blooming,
> The flower cares not for spring.
> The rich-presenting peony bursts late,

In early summer Shaoyao[20] luxuriates.
Only it never comes to a stop,
All the year round dancing.

Whatever dazzling beauty Peony and Shaoyao possess, their lives are transient. Yueji, instead, enjoys longevity. The poet Su eulogizes the flower's exuberant vitality explicitly.

Han Qi (1008 - 1075), a sage prime minister and renowned general in the Northern Song, also employs the technique of contrast to extol Yueji. His *The Chinese Rose in the Eastern Hall* goes:

The peony, clinging to the vernal breeze, is incommensurably charming and tender,
Whereas the dewed chrysanthemum is sparse with complaint in the gloaming cluster.
What else can rival this very floret in garishness and floridity,
With her Decennial efflorescence in delicate and deep rubicundity?[21]

Here, peony and chrysanthemum serve as two foils to Yueji, this perennial-blooming flower. In opposition to other flowers' short duration, the peerless Yueji is privileged with everlasting vitality. Probing more deeply, the Chinese rose here is a symbol which can be interpreted from a variety of perspectives.

Besides Yuejihua and Yueji, as a result of its long flowering, the Chinese rose has been crowned, in China, with such titles as Chang-chunhua (literally meaning "long-lived flower"), Sijihua (meaning "four-season flower"), Yueyuehong (referring to "monthly red"), etc. Two poems both titled *Changchunhua* are often anthologized as wonderful pieces on the Chinese rose. Not surprisingly, the two poets, one called Xu Ji (1028 1103), a deaf instructor and scholar, and the other, Zhu Shuzhen (c. 1135 c. 1180), a well-known woman poet, both lived in the Song Dynasty. The male poet remarks:

Who says God is impartial
Letting springtime reside solely in the flower?
Clouds are luster shining in your green leaves,
Petals are dyed red as the Sun's grant.
You, peach and plum's companion in the rain,

Witness of autumn sycamore's falling solitude,
Drink up your master's songs and wine,
And busy the flower-selling elderly.

Here, Yuejihua's unique trait is highlighted again but in a hyperbolic way— drinking while enjoying the flower and flower-selling recall the age when the poem was written.

Under the pen of the female poet, Changchunhua is delineated both directly and indirectly:

Fading – blooming, the flower is
Set by Spring Goddess bustling.
Gorgeous beauty peony claims,
Momentary duration dangles.

Setting the fully occupied Changchunhua against the gorgeous but leisurely peony, the lyrical poem is not lacking in wisdom or taste: diligence and persistence bring forth immortality.

The above-mentioned poet Yang Wanli left a rich legacy to posterity, among which a great number of poems on nature are still chanted and cherished even today. In whatever sense, he was always a staunch nature-lover, and his "Red Meigui" and "Yuejihua" are simply trustworthy demonstrations. The title "Red Meigui" is identical to Robert Burns' "A Red, Red Rose," but unlike Burns in likening his love to a red rose, the poet Yang describes the red rose straightforwardly:

Named differently from Yueji,
She is not like Qiangwei either.
Green leaves and stems intertwine,
Light and dark red petals overlap
Glamorous beauty dabbed in rouge,
She is the artistic creation of rain and dew.
Not attracted to other incenses,
The poet is left intoxicated.

If "Red Meigui" places greater stress on Meigui's sweet and rich perfume, then the poet's Yuejihua gives more prominence to the flower's perennial blooming.

When most flowers have a ten-day florescence,
This flower is bathed in the four seasons.
Near the burst one bud stretches,
Another blooms forth out of green stalks.
Transcending peach and plum's fragrance,
In snow and frost she combats plum blossoms.
Happily picking one to celebrate the New Year,
I forget it's de facto a winter morning.

Contrasting the Chinese rose with peaches, plum fruits and plum blossoms, the poem resounds with the poet's love and praise for the flower. Beauty and determination are inherent in the flower and hence, it deserves exquisite poetic lines.

The Chinese rose appealed to a bevy of poets in the Song Dynasty of China, which is a fact revealed naturally in the foregoing works. Moreover, in the Northern Song and the Southern Song, poetry creation centering on the Chinese rose came to culmination. However, the Chinese poets have never stopped making the flower one of their central poetical subjects. Three poems, all entitled "Yueji-hua," for instance, written respectively in the Ming era and Qing Dynasty, have been valued as representative works caroling the flower.

The Ming Dynasty poet Zhang Xin (?) sings:

Florescent wind breeds blossoms,
Newly-born dances with spring breeze, withering petals
falling into dust.
Only the flower rests not
Possessing oft the four seasons.

Liu Hui (?), another Ming poet, writes:

Green thorns lustrously grow,
Red buds burst one month after another.
Morning flowers draw out in the winding pond,
Evening petals load the stand.
Combating chrysanthemums under the frost,
She welcomes plum blossoms in the snow.
Dancing and singing on the green bank,
Once and again visitors are intoxicated.

The literary historian Sun Xingyan (1753 - 1818) in the Qing Dynasty says:

> With the snow plum, integrity is kept,
> Against peaches and plums the wonderful flower competes.
> Ladies-in-waiting come to enjoy,
> You are true beauty on the earth.

Following in the former poets' steps, the three poets likewise resort to the device of contrast. In borrowing such images – peaches and the snow plum – the Chinese rose's distinctive characteristics are presented to the full. Meantime, virtues have been planted in the flower, such as persistence, strong will, nobility, vitality, immortality and beauty. For this reason, the Chinese rose is a unique symbol permeated with cultural dimensions and extensions.

The "Chinese Rose"[22] and Classic Chinese Fiction

In actuality, the glamour of the Chinese rose can be seen not only in traditional Chinese poetry, but also be strongly felt in Chinese fiction, either in classical novels or in modern ones.

One of the Four Great Classical Novels of Chinese literature, *Journey to the West* is one reliable testimony. Written in the 16th century during the Ming Dynasty and attributed to Wu Cheng'en (c.1510 - 1582), the book in English-speaking countries is widely known as *Monkey: A Folk-Tale of China* based on a popular translation by Arthur Waley, an English orientalist and sinologist. In the 100 chapters of the fictionalized account of the Buddhist monk Xuanzang's legendary pilgrimage to India in the Tang Dynasty as well as an adaptation of traditional Chinese folk tales, the author Wu Cheng'en relates how the master Sanzang and his three disciples – Sun Wukong (Monkey), Zhu Bajie (Pigsy) and Sha Wujing (Sandy) – in search of sacred scriptures fight their way to the "Western Regions" despite innumerable dangers and toils. Rose, as an emblem of nature, also makes its appearance in this world classic.

For example, in Chapter 38 subtitled "Questioning His Mother, the Boy Sorts Right from Wrong/When Metal and Wood Join in the Mystery, Truth and Falsehood are Clear," when Monkey and Pigsy come to the Royal Gardens, they behold a desolate garden scene:

Carved and painted balustrades all in a mess,
Precious pavilions leaning awry.
The sedge and raspberries [have] been ruined.
Gone is the fragrance of rose and jasmine,
Tree peony and wild lily flower in vain.
Hibiscus and rose of Sharon are overgrown,
And all the precious choked.[23]

Chapter 82, "The Girl Seeks the Male/The Primal Deity Guards the Way" describes another garden:

East and west of the kiosks and balls are found
A wooden Mountain,
A Turquoise Screen Mountain,
A Howling Wind Mountain,
A Jade Mushroom Mountain,
All covered in phoenix-tail bamboo.
Trellises of briar roses,
And garden roses,
Growing by a swing,
As a curtain of silk and brocade.[24]

Again in Chapter 94 subtitled "The Four Monks Dine to Music in the Palace Gardens/One Demon Loves in Vain and Longs for Bliss":

The garden was a fine place:
Peony pavilions,
Rose bowers
Make a natural brocade.
Trellises of jasmine,
Beds of begonia,
All like sunset clouds or jeweled mosaics.[25]

Similarly, in Chapter 95, "False and True Form Combine When the Jade Rabbit is Captured/The True Female is Committed and Meets with Origin," a poem narrates the scene when Monkey says to the King that the Princess is a fake:

A roaring wind in spring,
The howling autumn gale.

When the roaring wind in spring blows through the wood
A thousand blossoms are shaken;
When the howling autumn gale hits the park
Ten thousand leaves all swirl and fly.
The tree peony beneath the balustrade is snapped;
Herbaceous peonies beside the balcony fall over.
Hibiscus on the pond's banks are shaken all about.
While chrysanthemums are flung in heaps at the foot of the
terrace.
The delicate begonia collapses in the dust,
The fragrant rose is now sleeping in the wild.[25]

And at the end of the novel, in Chapter 100, "This was indeed the great land of China, no ordinary place," just look:

Colored silks hung from the gates,
Red carpets were spread on the floor.
Heavy, rare fragrances,
Fresh and exotic foods.
...
Tender braised turnips,
Sugar-dredged taros,
Wonderful Sweet Mushrooms,
Fine fresh seaweed,
Several servings of bamboo shoots with ginger,
A number of rounds of mallows with honey,
What gluten with leaves of the tree of heaven,
Tree fungus and thin strips of bean curd,
Agar and aster,
Noodles with ferns and dried rose-petals,
Peppers stewed with radish,
...
Words could not describe the countless delicacies:
The great land of China was not the Western barbary.[26]

The illustration of the rose in *Monkey* attests that rose cultivation, in the Tang Dynasty, was already widespread, and one of the usages of the flower was introduced: the edible rose-petals could be mixed with noodles as food.

In the Qing Dynasty in Chinese history, about half a century after the publication of *Dream of the Red Chamber* by Cao Xueqin (1715 or 1724 - 1763 or 1764), one literary tour-de-force was in print, a fantasy novel titled *Jinghuayuan (Flowers in the Mirror,* also translated as *The Marriage of Flowers in the Mirror, Romance of Flowers in the Mirror)*. The 100-chapters of erudition, completed in 1827 by the novelist and phonologist Li Ruzhen (Li Ju-chen) (c. 1763 – c. 1830), is mainly about the protagonist Tang Ao's encounters with twelve incarnated flower-spirits during his exotic journey to strange lands, which is reminiscent of Jonathan Swift's *Gulliver's Travels* but meanwhile can be read as an encyclopedia of everything, resembling Melville's *Moby Dick.*

In the 5th chapter of this masterpiece, Shangkuan Waner, Imperial Concubine and personal secretary to China's first and only female emperor Wu Zetian, classifies 36 flowers into 3 categories: 12 teachers, 12 friends and 12 servants. When Princess Taiping says Yueji should be listed in the servant rank instead of being seen as a friend, Shangkuan answers back that Yueji repeats flowering in the four seasons, and thus it has a stable temperament, which has enabled it to be a worthy friend.[28]

Of course, Shangkuan spoke here in the person of the author himself. Li's appreciation of Yueji is revealed between the lines.

The rose is carried forward into 20th-century China which saw the birth and death of a famed woman writer known as Zhang Ailing (Eileen Chang) (1920 - 1995). Noted for her fiction that deals with the tensions between men and women in love, Eileen Chang, one of the greatest novelists of her generation, in her turbulent and tragic lifetime created many an impressive work, including *Love in a Fallen City, The Golden Cangue, Lust, Caution,* etc., and among them *Red Rose, White Rose,* an eloquent and evocative novella written in the 1940s, stands out too. The title of the novelette is reminiscent of the celebrated Wars of the Roses which lasted from 1455 to 1485 in English history. But differing from the red rose, the heraldic symbol of the House of Lancaster, and the white of the House of York, the two roses in this book refer to two women in the protagonist Zhenbao's life: one his spotless, chaste wife, the other his seductive, passionate mistress, who are, significantly, symbolized in ideal form – white rose and red rose respectively.

"Maybe every man has had two such women – at least two. Marry a red rose and eventually she'll be a mosquito-blood streak smeared on the wall, while the white one is 'moonlight in front of my bed.' Marry a white rose, and before long she'll be a grain of sticky rice that's gotten stuck to your clothes; the red one, by then, is a scarlet beauty mark over your heart."[29] The narrator's comment is, to be accurate, Eileen Chang's insightful findings and perceptions of man's mentality and the unequal status of the two sexes that she aims to convey to her reader.

Therefore, in contrast to the novelists in ancient China, the modern Chinese writers like Eileen Chang have enlarged the spectrum of the rose to its more profound implication and social bearing, which has beyond any inquiry and suspicion enriched the symbolic connotations of the rose.

The Chinese Rose and Chinese Folklore

Howard Giskin in his *Chinese Folktales* defines "Chinese folklore [as] include[ing] songs, poetry, dances, puppetry, and tales. It often tells stories of human nature, historical or legendary events, love, and the supernatural, or stories explaining natural phenomena and distinctive landmarks."[30]

Regarding religious and ideological influence on Chinese folk tales, the main ones have been "Taoism, Confucianism, and Buddism. Some folk tales may have arrived from India or West Asia along with Buddhism; others have no known western counterparts, but are widespread throughout East Asia."[31] The story of Qi Xi, also known as the "Story of the Magpie Bridge" or the "Story of Cowherd and the Weaving Maid," which tells how the stars Altair and Vega came to assume their places in the Galaxy, and the story of Hua Mulan, the female warrior who disguised herself as a man, are only two fine specimens of countless well-known Chinese folk tales. Relating to the Chinese rose, legends and folk tales are not inconsiderable either.

In Laizhou, Shandong Province, the noted cultivation base and hometown of the Chinese rose in China, for example, an interesting legend has been passed down from generation to generation. Local folklore has it that one day, in preparation to celebrate Queen Mother's birthday, the Chinese Rose Fairy specially picked a basket of the most beautiful Chinese roses. On her way to the Heavenly

Lake, the summer palace of the Queen, she passed by Laizhou. Finding it a pretty and attractive place, she could not help descending to the earth to enjoy the pleasant scenery. At the foot of Yufeng Mountain, a mountain in Laizhou, she met a handsome young man. When the Chinese Rose Fairy asked him the way, the man volunteered to be a tourist guide. Touched deeply by the young guy's hospitality and countrified simplicity, she fell in love with him. Further, when hearing that as a gardener, he loved and grew flowers, the Chinese Rose Fairy felt more joyful. While being immersed in delight and wonder, she suddenly remembered her task. In a hurry, she ran back to the flower basket without even saying farewell to the young man. However, to her surprise, the Chinese rose had already struck root and burgeoned forth. Having no other choice, she had to reach the Heavenly Lake empty-handed. Seeing the Chinese Rose Fairy carrying no roses with her, the Queen Mother questioned her closely. After being told the Chinese Rose Fairy had secretly been down to the secular world, in a rage, the Queen expelled her from the Southern Heavenly Gate and sent her into exile in Laizhou. Coming to Laizhou again, the Chinese Rose Fairy found the young man and married him. Thereafter, the loving couple planted flowers together. Under the Chinese Rose Fairy's cultivation, flowers, particularly the Chinese rose, became more dainty, gorgeous and fragrant and, with her help, more local people began to master the flower-planting skill. Henceforth, the Chinese rose in Laizhou became famous.

In addition to this tale, many other Chinese rose stories are available in rich Chinese folklore. These legends manifest fully that the Chinese rose has been a token bearing Chinese people's love and pursuit for a better future. Meanwhile, the influence of the Chinese rose on Chinese folk culture is self-evident.

The Chinese Rose as City Flower

The Chinese have always been enthusiastic in their unalterable love for flowers, and thanks to that love, China has so far bred a wide variety of incredibly rare flowers favored at home and abroad. In 1987, after 10 months of committee deliberations including 114 flower experts, results for the top 10 traditional Chinese flowers, determined by 149,000 votes from 29 different provinces, cities, and regions of China, were revealed. In order of preference, these are plum blossoms, peonies, chrysanthemums, orchids, Chinese roses, azaleas, camellias, lotus, osmanthus, and daffodils.

With the passage of time, Chinese people have granted abundant interpretations and cultural weight to the flowers they take much pride in and pay tribute to. Reputed for braving the harsh wintry days, the plum blossom takes the crown as the chief of the flower world. With a cultivation history dating back 4,000 years, the bloom is emblematic of unyielding will, fortitude and dignity in Chinese paintings and literature. First gaining avid interest in the Sui and Tang Dynasties, esteemed representing nobility, value, and peace, the exquisite peony has been called the "king of flowers."

The flower of Royalty – the chrysanthemum – has been steadily blossoming in China for 2,500 years, while the sage Confucius once said of the orchid, "[It] grows deep in the valley, even if no one is about, it emits its fragrance. / The junzi cultivates the Way and establishes his virtue; even if he is in difficulty or poverty, he does not waver in his integrity." Having a personal preference for the flower, Confucius equated the orchid with Junzi (a gentleman of noble character), both standing for perfection, nobility and virtue.

In Chinese culture, the azalea is known as the "thinking-of-home bush" (sixiang shu in pinyin) and is immortalized in the poetry of the great Tang Dynasty poet Tu Fu (712 770) symbolizing heroism and good luck. In contrast, the camellia, native to Sichuan and Yunnan and meaning, literally, tea flower, is valued for its blossoming in cold weather.

As to the lotus, Buddhists in the world recognize it as the holy seat of the Buddha in Buddhism. To the Chinese people, it is also a symbol of ultimate purity and perfection for it grows untainted out of the mud. With a long flowering period, the edible osmanthus flower is a mark of friendship, peace and good fortune.

And lastly, the daffodil is a water fairy pregnant with such virtues as nobility, purity and moral integrity.

Although ranked only fifth among the top ten, the Chinese rose is hailed as the queen of the flower kingdom. The national flower of the USA, Luxemburg, Cambodia, Morocco, Iraq and more, it is equally warmly accepted in its home country China. Not surprisingly, more than 50 Chinese cities – Beijing, Tientsin, Dalian, Shijiazhuang, Handan, Zhengzhou, Kaifeng, Anqing, Qingdao, Weihai, Jinzhou, Shashi, Jingzhou, Hengyang, Nanchang, Deyang, Xianyang,

Changzhi, Huaiyin , Taizhou, etc. – have adopted it as their city flower, which has made it China's top city flower. Its sister Meigui is also the choice of many Chinese cities, for instance, Shenyang, Lhasa, Lanzhou, Fushun, Yanji, Yinchuan, Chengde, Jiamusi and Urumqi. It was in 1987 that the Chinese rose, coupled with chrysanthemums, was elected Beijing's city flower, and as a result, Beijing, the capital of China, became the first Chinese city to enthrone the flower as its image ambassador.

A city flower is an important brand of a town's image and a calling card as well. The popularity of the Chinese rose in so many Chinese urban centers certifies the indisputable, unshakable status and the immense, far-reaching influence of the flower in Chinese culture; in the meantime, the fact has authenticated that the Chinese rose has maintained a solid and substantial foundation, beloved by the masses.

The Rose Industry in China

Though the flower has long been cultivated as an ornamental plant, as the native land and distribution center of the Chinese rose, China has known how to make use of the flower since ancient times.

Although in cuisines of today's Middle East, Turkey, Iran and North Africa, rose petals are widely used, it has been universally recognized that China was the first ancient country to use the rose as a spice. The flowers and fruits have a place as well in traditional Chinese medicine to treat irregular and/or painful menstruation as well as swollen thyroid. In health maintenance and beauty treatments, the flower is also prized as effective. The Empress Consort Wu,[32] the monarch of her self-proclaimed "Zhou Dynasty," was recorded to have drunk rose nectar in the morning and to have applied rose petals on her face in the evening, and therefore still looked young even in her sixties. Another powerful woman, Empress Dowager Tzu-hsi who controlled the Manchu Qing Dynasty from 1861 to her death in 1908, also bathed in rose water and beautified herself with rose petals. So the flowers worked wonders and age failed to wither her. As a matter of fact, in honest terms, the Chinese ancestors utilized the Chinese rose fully, and it is rather challenging to list all the usages here.

In the new age, China has, to its delight, witnessed the heritage maintaining of the traditional rose culture on the one hand, and on the other, the burgeoning of the rose industry. Pingyin County in Shandong and Kushui Town in Gansu have been respectively credited as the "National Rose Hometown of China" and the "First National Rose Town of China." Interestingly, there is indeed a town called Rose Town in Pingyin County. Pingyin, Kushui and Beijing's Miaofeng Mountain are the three known rose growing areas. In Pingyin and Kushui, rose festivals have been consecutively held in recent years and have further promoted China's rose industry. "In the Gully to the east of Miaofeng Mountain lies Beijing's famous wild rose valley. Each year beginning at the end of May, the slopes of this valley are festooned with countless rose blossoms."[33] Chinese rose festivals are celebrated in China too, and Tianjin's is one case. Held from May to June, Tianjin China Rose Festival has become the yearly Chinese rose celebration established in 1991, when people from all parts of the world can enjoy various flower exhibitions, operas and performances. On May 23, 2009, with the formal opening of the first Beijing Chinese Rose Cultural Festival at the Chinese Rose Park of Beijing Botanical Garden, the Chinese rose embraced its first festival in the capital of China.

The burgeoning development of the rose industry in China is one strong confirmation of Chinese people's valuing and maintaining of the traditional cultural heritage and at the same time one indication of their firm determination to carry forward this rich legacy.

To conclude, these innumerable details have proved one fact – the Chinese rose, together with its two Chinese sisters –, has deeply infiltrated into various aspects of Chinese culture and, consequently, has become an integral part of Chinese tradition. Born in Cathay, the flower, however, has also set its imprint on foreign lands and changed the rose world. Ergo, the unique contributions the Chinese rose has made will never be negated. It is no exaggeration to say that the Chinese rose has been a flower bridge connecting China with the outside world. Blossoming luxuriantly in the "Mother of Gardens," the Chinese rose truly deserves its title – the "Queen of Flowers."

Notes

[1] "Cathay" here refers to China in general, different from the ancient name for China.

[2] In ancient China, the Song Dynasty (960 - 1279) was divided into two distinct periods: the Northern Song (960 - 1127) and the Southern Song (1127 - 1279).

[3] Nova Beta. "Mother of Gardens." http://www.pbs.org/wgbh/nova/nature/china-

plants.html. Retrieved 8 July 2012.

[4] See "Inside Nova." http://www.pbs.org/wgbh/nova/insidenova/author/susan-k-lewis/. Retrieved 8 July 2012.

[5] Wikipedia. "Ernest Henry Wilson." http://en.wikipedia.org/wiki/Ernest_Henry_Wilson. Retrieved 9 July 2012.

[6] See Note 1.

[7] A book introducing plant cultivation in the Ming Dynasty of China, edited by Wang Xiangjin (1561 1653).

[8] China's first comprehensive and systematic encyclopedia on Chinese flora edited by Chen Junyu and Cheng Xuke, first published in 1990 by Shanghai Culture Press, PRC.

[9] A book on the study and practice of growing plants.

[10] The other three great poets at the time were: Lu You (1125 - 1210), Fan Chengda (1126 - 1193) and You Mao (1127 - 1194).

[11] One of the greatest poets of the Tang Dynasty, China's "golden age" of classical Chinese poetry, known as the Poet Sage whose ideas had a profound impact on American Imagist and Modernist poetry.

[12] A book on flowers by the Qing Dynasty horticulturist Chen Haozi (c.1612 ?).

[13] The first book on plant illustration in China by Wu Qijun (1789 - 1847), a Qing Dynasty scholar and politician.

[14] Q. Wang et al. "Climatic Change during the Palaeocene to Ecocene Based on Fossil Plants from Fushun, China." http://sourcedb.ib.cas.cn/cn/ibthesis/201009/P020100919552692017883.pdf. Retrieved 8 July 2012.

[15] All-America Rose Selections. "Overview." http://www.rose.org/site/epage/65027_766.html. (8 July, 2012).

[16] The name literally means "Divine Farmer" who was a legendary ruler of China and culture hero living some 5,000 years ago.

[17] The ancient trade routes that spanned Asia, Africa and Europe from the Han Dynasty of China until about 1,400.

[18] An English horticulturalist, artist, author, poet and garden designer who was best known for his work in gathering and popularizing old and new shrub roses.

[19] A type of classical Chinese literature between poetry and prose, originating in the Tang Dynasty of China and fully developed in the Song Dynasty.

[20] Chinese pinyin referring to Chinese herbaceous peony.

[21] "The Chinese Rose in the Eastern Hall." http://www.njyl.com/article/s/581094-323517-0.html. Retrieved 30 June 2014.

[22] Three classic Chinese novels are particularly focused on in this section: *Journey to the West*, *Flowers in the Mirror*, and *Red Rose, White Rose*; but in the first and the third works, Meigui (the rose) and Qiangwei are referred to, while in the second writing, Yueji (the Chinese rose) is alluded to. Quotation marks are employed for this case.

[23] Cheng-en, Wu. *The Journey to the West*. Trans. Anthony C. Yu. Chicago: University of Chicago Press, 1977, 535. http://www.qcenglish.com/ebook/2796.html. Retrieved 8 July 2012.

[24] Ibid., 1155.

[25] Ibid., 1313.

[26] Ibid., 1324-25.

27 Ibid., 1396-97.

28 See the translated English version of the book: Ju-chen, Li. *Flowers in the Mirror*. Trans. Lin Tai-yi. California: University of California Press, 1965.

29 Eileen, Chang. *Love in a Fallen City and Other Stories*. Trans. Karen S. Kingsbury and Eileen Chang. NY: Penguin Books, 2007. 255.

30 Giskin, Howard. *Chinese Folktales*. Chicago: NTC Publishing Group, 1997. http://en.wikipedia.org/wiki/Chinese_folklore. Retrieved 20 July 2012.

31 Wolfram, Eberhard. *Folktales of China*. Chicago: University of Chicago Press, 1965. http://en.wikipedia.org/wiki/Chinese_folklore. Retrieved 10 July 2012.

32 A dynasty which existed from 690 to 705, established by China's first and last female monarch Wu Zetian.

33 Visit Beijing. "Miaofeng Mountain and Wild Rose Valley." http://beijing.english.china.org.cn/2008-05/17/content_15287283.html. Retrieved 10 July 2012.

References:

Cheng-en, Wu. *The Journey to the West*. Anthony C. Yu, Trans. Chicago: University of Chicago Press, 1977.

Eileen, Chang. *Love in a Fallen City and Other Stories*. Karen S. Kingsbury and Eileen Chang, Trans. New York: Penguin Books, 2007.

Ju-chen, Li. *Flowers in the Mirror*. Lin Tai-yi, Trans. California: University of California Press, 1965.

Boyun, Sun (孙伯筠) and Chi, Zhang (张持). *Huajiandao: Huamu wenhua jianshang* (花间道：花木文化鉴赏*Appreciating Flower and Tree Culture*). Beijing: China Agriculture Press (中国农业出版社), 2008.

Ce, Chen (陈策). *Meigui jianshang yu wenhua* (玫瑰文化鉴赏与文化*The Rose Appreciation and Culture*). Guangzhou: Guangdong Science and Technology Press (广东科技出版社), 2008.

Guanghong, Yin (殷广鸿). *Gongyuan changjian huamu shibie yu jianshang* (公园常见花木识别与鉴赏*Identification and Appreciation of the Common Flowers and Trees in Parks*). Beijing: China Agriculture Press (中国农业出版社), 2010.

Shuxun, Yu (余树勋). *Yueji* (月季*The Chinese Rose*). Beijing: Jindun Publishing House (金盾出版社), 2008.

Shiliang, Gao (高世良). *Baihua baihua* (百花百话 *Flowers and Messages*). Tianjin: Baihua Literature And Art Publishing House (百花文艺出版社), 2007.

Wen, Fan (范文). *Xiangyan pinwei baihua* (香艳品味百花 *Flower Appreciation*). Shenzhen: The Rock House Publishers (海天出版社), 2007.

Yongmei, Li (李永梅). *Huayan qiaoyu* (花言巧语*Flowers and Their Symbolic Meanings*). Tianjin: Tianjin Ancient Books Publishing House (天津古籍出版社), 2007.

Yu, Chen (陈裕) and Xiaofei, Luo (罗小飞) et al. *Zhongguo minghua jianshang* (中国名花鉴赏 *Appreciation of the Famous Chinese Flowers*). Beijing: China Architecture and Building Press (中国建筑工业出版社), 2010.

Zuzhang, Jia (贾祖璋). *Huaniao chongyu yu wenxue* (花鸟虫鱼与文学 *Flower, Bird, Insect, Fish and Literature*). Changsha: Hunan Educational Publishing House (湖南教育出版社), 2002.

11

A Survey of Rose Symbolism in Traditional Irish Music

By Ryan Dunne

My first foray into Irish music began during my final year of high school in Fairfax, Virginia, a town just outside Washington, D.C. The girl I fancied at the time approached me and exclaimed, "Oh my gosh, you are so Irish. I love your soul!" For a moment, I couldn't imagine how she could tell, but then I realized I was wearing my Claddagh ring and necklace. And a tweed hat, of course, as the weather demanded it. To be honest, my sweatshirt may have had an Ireland logo, too. Well, clearly I was fond of Ireland, but she was the first person to suggest that my ancestry might also be connected to my soul. What's an Irish-American boy to do when he wants to win a girl's heart and find his soul? I picked up the tin whistle and learned a few Irish tunes. "Tis the Last Rose of Summer" was one of the first tunes I learned and performed for her.

Though in the end I failed to win her heart, Ireland had won mine. My first trip to the mother country as a shy college student took me to a small family farm in County Meath. During their family reunion, my host family asked me to play a tune on my tin whistle. Up until that point I knew how to play three tunes and had only one memorized well enough to play. So I shakily took a swing at "Down By the Sally Gardens," the first song in the *How to Learn Irish Tin Whistle* book that came with my instrument. To my amazement, this family of twelve or so all began to sing along. I didn't even know it *had* lyrics, and they turned the two traditional verses into a dozen.

Leon Uris's novel *Trinity* first made me aware that parts of Irish culture are in danger of being lost. Further reading exposed me to England's forced dismantling of Irish culture throughout the ages. The penal laws in the 17th century were the harshest, making it illegal for the Irish to speak their own language, teach their children, or practice their Catholic religion.[1] I made it my purpose to learn about the Irish language, culture, and music. I travelled and studied in Ireland and learned more songs, met more people, even discovered dozens of extended family members. My father's side is from Dungarvan, County Waterford, an area famous for the now-shuttered Waterford Crystal. My mother's side is from County Mayo, home to the holy mountain Croagh Patrick. My wife and I conquered this mountain in a painful five hours, only to learn that an extended cousin had recently climbed up and down it twelve times (a cumulative 18,134 meters) within a single day to beat a world record for the most meters climbed in 24 hours. I learned that my grandparents' favorite singers were the famed Irish tenors John McCormack (1884 - 1945) and Frank Patterson (1938 - 2000), and strove to learn the songs they had sung and their significance in Irish history.

But I still didn't truly understand Irish music until I discovered the rose. I found it soon after my art school graduation, when I bought *The Grand Irish Songbook.* It was in songs such as "The Rose of Tralee" and "My Wild Irish Rose," and I began to sing these songs to my girlfriend at the time. I won't tell you if they worked, but I will tell you it takes her five hours to climb Croagh Patrick. I have since learned and performed every Irish rose song I can find, dating from the 16th century all the way to the present day. You can probably guess that when Frankie Hutton introduced me to The Rose Project, I couldn't possibly say no to writing this essay. Since then, I've been researching and categorizing the symbolism of roses in Irish folk songs and discovered that almost all mentions of the flower fall into one of three categories: first, as a metaphor for the beauty of a maiden; second, as a political symbol marking significant points in history; and third, as a symbol for the passage of life and death.

Comparing the beauty of a fair lady to a rose is by far the most common function of the rose in Irish music. Though no one has ever taken a Rose Census in the field, you don't even have to go past the title to find a nearly endless list of Irish folk songs praising youthful maidens as roses, e.g. "My Rose of the Mountain" (~1972), "The Ringsend Rose" (1970s), "My Wild Irish Rose" (1899), "Sweet Rosie O'Grady" (1896), "Red is the Rose" (1841) to name a few.

And if you start listing songs with "The Rose of ____" in the title, I wouldn't be surprised if there was a song for every county in Ireland: The Rose of ... Tralee, Allendale, Avondale, Mooncoin, Castlerea,

Killarney, Kilkenny, Aranmore, Connemara, Clare, and this list is limited to the 19th and 20th centuries.

These are all songs that one way or another compare the beauty of a rose to a woman. Of the three basic categories of rose songs, I have found these to be the simplest and sweetest. A line from "Sweet Rosie O'Grady" explains,

> Her name is Rose O'Grady and,
> I don't mind telling you,
> That she's the sweetest little Rose
> the garden ever grew.[2]

Nothing complicated here; this is a direct comparison of something beautiful to the woman of one's desires. A line from "The Rose of Allendale" concurs:

> Though flowers decked the mountainside
> And fragrance filled the vale
> By far the sweetest flower there
> Was the Rose of Allendale.[3]

The sweetest flower of them all in Allendale is the Rose. A line from "The Rose of Mooncoin" agrees:

> Flow on, lovely river, flow gently along
> By your waters so sweet sounds the lark's merry song
> On your green banks I wander where first I did join
> With you, lovely Molly, the rose of Mooncoin.[4]

The locations change, but the metaphor stays the same. Let's take a look at the first verse and chorus of the song "Red is the Rose" (~1841):

> Come over the hills, my bonnie Irish lass
> Come over the hills to your darling
> You choose the rose, love, and I'll make the vow
> And I'll be your true love forever.
> Red is the rose that in yonder garden grows
> Fair is the lily of the valley

Clear is the water that flows from the Boyne
But my love is fairer than any.[5]

Now, apart from using the rose to symbolize a vow of true love, the song also tells us that the speaker's true love is fair, even more so than the rose or the lily. This swoon-inducing comparison has been around since Biblical times, as seen in the "Song of Solomon," second chapter, first verse:

I am the rose of Sharon, and the lily of the valleys.
As the lily among thorns, so is my love among the daughters.[6]

This shows that rose and lily metaphors were already ancient before Dublin even existed! Perhaps this passage inspired the use of the rose in Irish music; the similarity is remarkable, and a people whose monks preserved the Bible through the Dark Ages[7] can certainly be expected to have read it too.

As influential as scripture undoubtedly was, no one song can be responsible for the love affair the Irish imagination has had with the rose. Who can ignore the impact of the most famous Irish rose song of all, "The Rose of Tralee?" Is there another that has been celebrated with its own international festival for over half a century? "The Rose of Tralee" was written in 1845, and the lyrics are credited to Edward Mordaunt Spencer, the music to Charles William Glover. However, popular legend attributes the song to William Mulchinok, claiming that it recounts a real relationship that occurred during the famine years of Ireland around 1840.[8] Whether real or imagined, the powerful romance in the song made it an enduring fixture of Irish music, and in 2013, I traced its deep tracks to the town of its origin. It was there in Tralee that I discovered the song's festival, which is centered around an international beauty pageant featuring "Roses" (women) from countries across the globe. I was also lucky enough to visit Tralee's magnificent rose garden, another legacy of the song.

So, let's take a look at what all the fuss is about. Here's the chorus from "The Rose of Tralee":

She was lovely and fair as the rose of the summer,
Yet 'twas not her beauty alone that won me;

Oh no, 'twas the truth in her eyes ever dawning,
That made me love Mary, the Rose of Tralee.[9]

Again, we find a simple comparison. Irish culture certainly thinks highly of the beauty of the rose. But it also looks deeper, into "the truth," the purity of the symbol. This can be seen as well in how the Rose of Tralee International Festival considers itself more than just a beauty pageant. The festival website explains:

> A Rose reflects the intelligence, compassion and independence of modern Irish women. Over the years our Roses have mirrored a changing Ireland and the definition of being Irish that is celebrated by so many different people around the world. Roses have come to reflect the widening of Irish borders and the embracing of our global diaspora in a positive and refreshing way. A Rose represents the collective aspirations, social responsibilities and ambitions of young women from a variety of communities and backgrounds, united by their desire to celebrate their Irish heritage.[10]

This is by no means the only meaning the rose has to the Irish. It certainly shares a little with older symbolism. The Irish Gaelic song "Róisín Dubh" (Little Dark Rose), written in the 16th century, initially compared the dark rose to a beautiful aristocratic woman waiting for her love to come home from overseas, a common theme in Irish music. It was translated and given new meaning a number of times. The most notable version is by James Clarence Mangan under the title "My Dark Rosaleen."[11] Mangan's version represents the rose/woman as a "virgin flower" that is betrayed and divided by the masculine England.[12] A number of other poets have also been inspired by "Róisín Dubh." Take a look at the last line of Joseph Plunkett's poem "The Little Black Rose at Last Turns Red":

> Praise God if this my blood fulfils the doom
> When you, dark rose, shall redden into bloom.[13]

This phrase dangerously alludes to the fact that there must be bloodshed for the dark rose to bloom once again in freedom. Indeed, 19th century poetics often used the rose as a symbol of renewal through blood, crucifixion, sacrifice and spring.[14] During this era, speaking out openly against England could have dire consequences, and thus the continued use of the rose as a tacit symbol for Ireland was common.[15] The last verse of Mangan's "My Dark Rosaleen" reflects this beautifully:

> O, the Erne shall run red,
>   With redundance of blood,
> The earth shall rock beneath our tread,
>   And flames wrap hill and wood,
> And gun-peal and slogan-cry
>   Wake many a glen serene,
> Ere you shall fade, ere you shall die,
>   My Dark Rosaleen!
> My own Rosaleen!
>   The Judgement Hour must first be nigh,
> Ere you can fade, ere you can die,
>   My Dark Rosaleen![16]

"Róisín Dubh" inspired the hugely popular patriotic film *Mise Eire* (I am Ireland) and appears in the soundtrack.[17] It brought popular attention (and profit) to Irish musicians throughout the 1960s.[18] The song has even been taken on by more contemporary musicians such as Sinead O'Connor, Mary Black, and Thin Lizzy. Thin Lizzy's rock version begins with

> Tell me the legends of long ago
> When the kings and queens would dance in the realm of the Black Rose
> Play me the melodies I want to know
> So I can teach my children, oh …

and goes on to sing through countless well-known Irish melodies of old in the hope of passing them on to the next generation. The poet Joan McBreen explained in a 1992 interview:

> The influences are powerful even though we do not discuss this very much, if at all. For example, songs like "Cill Chais" ("Kilcash") and "Róisín Dubh" ("Little Black Rose") are still

sung and are greatly loved. I feel the rhythms in my poetry and in the poetry of my contemporaries, male and female.[19]

Hints of the black rose date back to ancient Ireland too. According to this curious legend told by Theo Stoof there lived a council of druids in Ireland near Briug na Bóinde.

> These Druids wore black and red robes on which their symbol, the Róisín Dubh, the Black Rose, was visible. They lived in a small mansion which none other than they ever entered. At night, an eerie light which slowly varied from one unnatural colour to another shone through the few windows of the house, but this never flickered like that of a candle or fire. Sometimes loud noises such as supernatural songs and ecstatic screams were heard coming from the mansion, and in surrounding villages it was rumoured that the Druids feasted upon the bodies of the dead at night.[20]

It's difficult, isn't it, to read about how a single song has influenced poets, musicians, film-makers, artists, and revolutionaries over the past 400 years without wanting to play it or hear it played? And there are others like it.

The song "The Bonny Bunch of Roses, O" (1881) imagines a conversation between young Napoleon Bonaparte II and his mother Joséphine concerning "a bonny bunch of roses." The son then promises to go out to raise a mighty army and win the roses for his mother. His mother derides him for thinking as foolishly as his father Napoleon who was exiled and died in the Isle of St. Helena because of this mission for roses. "The bonny bunch of roses" mentioned in the song is often said to represent the countries of Europe that Napoleon had been conquering throughout the early 1800's, but further research has led me to another theory. Napoleon's second wife, Joséphine de Beauharnais was also known as the Patroness of Roses. Joséphine had become obsessed with the rose and, after buying an estate and creating her own rose garden, demanded to have roses from all over the world collected and brought there. Napoleon was known to have sent warships to other countries to collect particularly rare roses for his wife.[21] With this in mind, I'd propose that the roses in the song represent the hearts of the countries that Napoleon conquered. Alternately, from a child's perspective, innocent young Napoleon II may have thought that his father's great campaigns to conquer the world were merely journeys to collect roses for his mother Joséphine.

The final major symbolic deployment of the rose we'll examine uses the flower to reflect upon the passing of life and death. In "'Tis' the Last Rose of Summer," prolific Irish poet Thomas Moore in 1805 does just that while also continuing the tradition of comparing a woman to the rose.

> 'Tis the last rose of summer,
> Left blooming alone;
> All her lovely companions
> Are faded and gone;
> No flower of her kindred,
> No rosebud is nigh,
> To reflect back her blushes,
> Or give sigh for sigh.
>
> I'll not leave thee, thou lone one!
> To pine on the stem;
> Since the lovely are sleeping,
> Go, sleep thou with them.
> Thus kindly I scatter,
> Thy leaves o'er the bed,
> Where thy mates of the garden
> Lie scentless and dead.[22]

The poem "A Rose will Fade" written by Dora Sigerson in 1893 represents the passage from life into death in a similar fashion.

> Why did you smile to his face, red Rose,
> As he whistled across your way?
> And all the world went mad for you,
> All the world it knelt to woo.
> A rose will bloom in a day.
>
> I gather your petals, Rose - red Rose,
> The petals he threw away.

And all the world derided you;
Ah! the world, how well it knew
A rose will fade in a day.[23]

Both speak of the beauty of a living rose and its purity but go on
to bemoan how a rose can wilt so quickly. Another poem by Dora
Sigerson, "Little White Rose" (1893), describes this transformation
very well. The poem begins rejoicing in gladness for the purity and
youthful blooming of the speaker's white rose. The rose in the poem
then grows red and the verse fills with talk of passionate tears and
love. The final verse marks the transformation from red to black as
the rose fades away from her love and gets lost in the alluring fire of
the world.[24] Another song by Thomas Moore, "Farewell, But Whenever You Welcome the Hour" contemplates the memories of friends
and a life and love gone by:

Long, long be my heart with such memories fill'd,
Like the vase in which roses have once been distill'd.
You may break, you may ruin the vase if you will,
But the scent of the roses will hang 'round it still.[25]

Even the cheery 1899 song "My Wild Irish Rose" by Chauncey
Olcott begins with a rose, faded and dead, that one must reflect and
think upon:

If you listen I'll sing you a sweet little song
Of a flower that's now drooped and dead,
Yet dearer to me, yes than all of its mates,
Though each holds aloft its proud head.
Twas given to me by a girl that I know,
Since we've met, faith I've known no repose.
She is dearer by far than the world's brightest star,
And I call her my wild Irish Rose.[26]

And the famous "Rose of Tralee" has a third verse that is rarely
sung where the speaker returns from a trip to India only to find his
precious rose had died during the Great Famine. While the rose is
often used to describe one's hope, love and passion, it has a dark
side as well.

Irish music is drawn deep from the soul of the Irish people, and when an old song is lost, it cannot be replaced. There is a profound depth to these songs, one that allowed a simple 16th century love song like the "Róisín Dubh" to be rewritten multiple times over the centuries, each time with a meaning as vital as the last. The changes reflect the political environment of the times while still staying true to the feeling of the ancient melody.

The songs spoken of here are only the beginning. Much more research must be done, for the rose is a resilient symbol, able to represent love and loss, life and death. Whatever happens, roses will surely continue to help Ireland speak its heart through music, now and forever. But however beautiful the blooms of the present may be, the singer of the Rose of Tralee can't forget about the bloom of the past, and neither can I. The rose songs of old inspire artists and musicians like me to spread, perform, and preserve them. Maybe, if you get a chance, you'd like to listen.

References

[1] Burton, Edwin, Edward D'Alton, and Jarvis Kelley. "Penal Laws." *The Catholic Encyclopedia*. Vol. 11. New York: Robert Appleton Company, 1911. np.

[2] Alfred Publishing Staff. *Great Songs of the Century: 51 Lucky Irish Classics*. NP: Alfred Publishing, 1992. 64.

[3] Leonard, Hal. *The Grand Irish Songbook*. Milwaukee: Hal Leonard, 2006. 260-263.

[4] Ibid. 264-266.

[5] Ibid. 239-243.

[6] "Song of Solomon." *The Bible*. New Testament, King James Version. New York: American Bible Society, nd.

[7] Cahill, Thomas. *How the Irish Saved Civilization*. NY: Doubleday, 1995.

[8] Locke, Peter; Ryle, Lilly; Browne, Eamon; Corkery, Tim. Interview by Maurice O'Keefe, audio book. *The True Story of the Rose of Tralee*. Kerry, Ireland, August 2004.

[9] *The Grand Irish Songbook*. Ibid. 267-269.

[10] "Kilkenny Rose Centre." Festivals in Ireland, Kerry, Tralee. January 1, 2014. http://www.roseoftralee.ie/community/userprofile/kilkennyrose Retrieved 8 November 2014.

[11] McPeake, Margaret. "Transformations in the Representation of the Nation as Woman." *New Voices in Irish Criticism*, 2000. 229.

[12] Kearney, Richard. *Navigations: Collected Irish Essays, 1976-2006*. Syracuse, NY: Syracuse UP, 2006. 8.

[13] Murphy, Maureen O'Rourke. *Irish Literature - A Reader*. Syracuse: Syracuse UP, 1911. 240.

[14] Kearney, Richard. *Navigations: Collected Irish Essays*, 1976-2006. Syracuse, NY: Syracuse UP, 2006. 37.

[15] Ibid. 38.

[16] Colum, Padraic, ed. *Anthology of Irish Verse*. NY: Boni and Liveright, 1922. 129.

[17] Rockett, Kevin, Luke Gibbons, and John Hill. *Cinema and Ireland*. London: Routledge, 1988. 86.

[18] Crosson, Seán. "Traditional Music and Song and the Poetry of Thomas Kinsella." *Nordic Irish Studies* 7 (2008): 105.

[19] Brandes, Rand. "An Interview with Joan McBreen." *Colby Quarterly* 28, no. 4 (1992): 260-64.

[20] Thomas, Graham Stuart. *The Graham Stuart Thomas Rose Book*. Enl. and Thoroughly Rev. ed. London: Frances Lincoln, 2004.

[21] See Stoof, Theo. "The Legend of Róisín Dubh." Ireland List - Irish Legends. http://freepages.genealogy.rootsweb.ancestry.com/~irelandlist/myth.html Retrieved 8 November 2014.

[22] Alfred Publishing Staff. *Great Songs of the Century: 51 Lucky Irish Classics*. Alfred Publishing, 1992. 75.

[23] Sigerson Porter, Dora. *The Collected Poems of Dora Sigerson Porter*. London: Hodder and Stoughton, 1907. 214.

[24] Ibid. 143.

[25] Moore, Thomas. *Irish Melodies*. Vol. 5. London and NY: Boosey & Co., 1895. 112. http://www.musicanet.org/robokopp/moore.html Retrieved 27 December 2014.

[26] Alfred Publishing Staff. *Great Songs of the Century: 51 Lucky Irish Classics*. NP: Alfred Publishing, 1992. 56.

12

"Dying, Laughing"
The Rose From Yeats to Rumi

By Frankie Hutton

We lovers laugh to hear "This should be more that and that should be more this" coming from people sitting in a wagon tilted in a ditch. Going in search of the heart, I found a huge rose under my feet, and roses under all our feet…

Rumi
(Trans. Coleman Barks)

This essay surveys timeless, metaspiritual qualities of the rose as reflected in classical poetry and literature. Serving as didactic guideposts, the renderings considered are no stranger to sophisticated academic communities, yet release only with reluctance their recondite meanings. Transcending vast epochs and geographies these passages teach esoteric knowledge to seekers who are awake and aware. Thus, from important writers of diverse eras and cultures, we encounter the same messages in different format and creative style.

To provide insight into higher consciousness, these authors, metaphysically astute, have sown the rose as a potent symbol in short stories and poems, knowing well that the flower provides an exceptional image 'planted' on earth to teach all who are ready about real life, death, and higher spiritual realms. Thus, unlike human guides who appear at strategic times when one is on the path, the rose is ever present, ever ready, ever useful and ever beautiful for those wise enough to attend to it.

Intriguingly, although no one knows exactly when roses first appeared, the lovely flower has nonetheless shown considerable staying power and clout, not unlike the cross, the oceans, the sun and the mountains. Eons ago, the quintessential rose took its place beside entities that far overstep the material mind to reach toward the ineffable and everlasting. Explored in this essay as a profound poetic metaphor, the rose easily and pervasively transcends what is known of its potent botanical qualities to extend deep into the mysteries of the universe and human existence. For by reading timeless literature and poetry connected with this magnificent flower, we find its full splendor revealed. Careful review of rose-embracing works reveals that the flower in all its perfection links directly to the Source of All and Everything. For those who have eyes to see, it comes into focus as one of the most long-lasting, widely-used and elegant, powerful symbols of all time.

Why else would we find its likeness all around us? For eons, the beauty and medicinal qualities of the rose have made it a cherished icon of innumerable organizations, clubs,[1] and municipalities,[2] as well as a powerful metaphor in movies,[3] dance,[4] fiction and non-fiction.[5] Barbara Seward's study of the rose in British, Irish and American masterpieces is used pivotally here to reinforce the notion of the flower's broad appearance in popular literature. Seward also offers allegorical and metaphysical readings of selections that transcend the material world.

In *The Symbolic Rose* Seward privileges Dante Alighiere's *Divine Comedy* as the primal literary antecedent to understanding the rose in William Butler Yeats (1865-1939), T. S. Eliot (1888-1965) and James Joyce (1882-1941). According to Seward, before Dante's *Divine Comedy* "no earlier writer had attempted to express through a single rose the diversity of meanings associated with" the flower.[6] Furthermore, no one before the Florentine "had demonstrated the poetic genius" or fully reconciled "such seemingly disparate qualities as carnal and spiritual love." While Seward's admiration is not misplaced, she could have reached even farther back into the globe's chronology to find the rose in mystically-fired literature long before Yeats and even Dante. What's more, Yeats himself seems aware of the rose's long-lived connection to the Divine for he often wrote about roses and divinity as corollaries.

In the *Symbolic Rose*, Seward introduces Yeats's foundation in nineteenth century theosophy and secret societies, while Eliot and Joyce "acquire a good part of their meaning from associations with the rose of Dante..." – that is to say from 14th century Italy. And even though interpreting the *Commedia* was challenging, all three authors were influenced by Dante's masterpiece, taught in the academy for centuries. The key they took from it is best expressed in Helen Luke's words. Dante's journey moves toward a "vision of God" and the single truth that "without divinity there can be no conscious humanity, and without humanity the divine remains an abstraction,"[7] a point I underscore in this and in the two essays following.

How does the *Divine Comedy* feature the rose? A short summary will show the epic poem's connection to the flower symbol. Under the influence of the Roman Church, Dante faced a profound inner struggle that led to exile from Florence between 1307 and 1321, the years when the *Divine Comedy* was written. The exile, simply put, resulted from the era's disputatious civic and church politics involving the Guelphs and the Ghibellines who split over investiture in a power struggle between the Pope and the Holy Roman Emperor. The ruling party, the Guelphs, was supported by Dante's family but split over the nature and power of the papacy. This divide resulted in a lasting feud in which Dante protested papal policies. Battles between the two groups continued throughout the decade and a half of Dante's exile.

Dante opens his epic saga during the Easter Season on Good Friday in 1300. The verses feature a traveler, interpreted as the author's persona, who encounters characters and symbols related to salvation and damnation. Significantly, the entire work is based on the number three – 3 –, an important cipher in metaphysics that suggests human beings must rise above this world's dualities to reach a higher place sans tension, fighting and foolishness. Thus, in sacred geometry, when a third force is introduced, two initial forces are neutralized. On his journey Dante is "turned back by three beasts – the leopard, the lion and the wolf – by his love of pleasure, by his fierce pride, and by the terrifying latent greed and avarice of the ego."[8] These are "beasts" that most of us must conquer in the material realm, in life, to move up and on to the Divine.

Not only images but also the poem's structure rely on three. Verses in the *Divine Comedy* are arranged in units of three lines; and the epic is composed of three books, *Inferno, Purgatory* and *Paradise*, all phases Dante encounters on his spiritual journey.[9] Stiff moral judgments are forced on Dante as he reaches Earthly Paradise. In paradise, Dante encounters Beatrice, the sacred feminine, his guide and inspiration. It is she who leads him toward God, revealed in "an ineffable vision of light."[10] Of course this heaven, a beautiful place, reveals the "perfect society mov[ing] in the harmony of the dance or form[ing] the flawless pattern of the rose."[11] It's easy to follow Dante's symbolism here, to imagine the rose as impeccable and faultless, in a word, prefect.

Now, the rose appears only toward the end of the epic. Why did Dante wait to introduce it? "Heaven," like the rose itself, represents the perfection for which we must strive as we work toward conquering our material 'beasts'. Overcoming earth's dualities to introduce the third force is not work for the faint of heart; it's daily striving performed in earnest all the time, twenty-four-seven in the vernacular. And, depending on the progress made this time around, the effort may continue in other incarnations as well. In Canto XXXII, Dante spells out, in seven steps or gradations, the prerequisite struggles essential to reach paradise, leaf by leaf as "throned on the rose." In this last Canto, Dante reveals something profound, veiled for millennia and yet accessible globally in many sacred books, the most prominent being the Christian *Holy Bible* but also significant fiction. In other words, the pathway is mapped for those who dare to undertake the journey through incessant work consisting of unconditional love, selfless service and untiring desire to know oneself and truth.

As for the guidepost, the rose continues as a marker on the upward path. Like beneficiaries of traditions featuring meditation and metaphysics, the initiated know that the phrase "leaf by leaf" alludes to the consecrated, unremitting work each individual performs to climb the full number of rounds and connect at the summit, fully and everlastingly, with the Holy Spirit (as Christianity names it). This process is described in sacred literature such as in the *Holy Bible*, but it is also deeply hermetic knowledge easy to miss for the spiritually uninspired. Dante, however, was inspired and wanted very much to be illumined. For as both writer and protagonist of his *Commedia*, he accentuates the pain and the beasts of life that we all encounter. His earthly or material body died in Ravenna in 1321 while away from his beloved Florence, and although we can't know whether or not he made the full journey to eternal life or endlessness, his poem shows an intimate acquaintance with the path and its impediments to reaching the Light. His epic continues to be a staple in world literature,[12] religious studies and philosophy. And no wonder. As Peter Bondanella's analysis shows, in confronting the seven deadly sins of lust, gluttony, avarice, sloth, wrath, envy and pride in purgatory, Dante "was under the influence of Saint Jerome's Latin Vulgate version of the Holy Bible from which he draws 600 references [in] the epic poem."[13]

Dante influences Yeats, Eliot and Joyce

Not unlike Dante's rose representing struggle against such 'deadly sins', Yeats's metaphysical connections tip us off to his use of the rose as a metaphor for attaining full initiation. In Yeats, the rose appears for instance in "To the Rose Upon the Rood of Time," "The Rose of the World," and "The Rose of Peace." The floral mission is revealed quietly in allegory, metaphor and symbol throughout much of Yeats's work. The question is why? Yeats, who attained considerable insight into mysticism, seems to want readers to perceive his journey and, as important, to feel that they too, can achieve similar spiritual growth even if, of the three great authors, Yeats's use of the rose appears the most esoteric.

To explore the flower in Yeats's contemporaries, we turn to Seward who offers a fine analysis of the image in the Anglo-American and the Irishman's oeuvre. Regarding T. S. Eliot, she shows that the rose is the "obvious symbol of whole and enduring resolution" even as it combines "the romanticism of a yearning, nostalgic, insatiable age with absolute, authoritarian standards of medieval times …"[14] As for Joyce, he also made full use of the rose in his prose and verse.

Yeats, however, and his poignant use of the rose in one particular short story, forms this essay's centerpiece, as Seward, too, considers his work a marker in the literary history of the rose. Using a "medley of artistic and religious concepts," Yeats makes the first successful attempt since Dante, in Seward's view, to express tradition and personal meanings in the single symbol of the rose.[15] Having joined the Dublin Theosophical Society in 1886, he then became a member of the Order of the Golden Dawn in London, a highly intellectual, hermetic fraternity,[16] and his initiation into the latter, an offshoot of the Rosicrucians, would have been equally profound. As the influence of these groups on his writing is undeniable, both played a role in Yeats' spiritual breakthrough. Although we don't know exactly when Yeats felt the full impact of their influence, it is noteworthy that the Theosophical Society, founded by H. S. Olcott[17] and H. Blavatsky, was launched in 1875. In 1890, the year before she died, Blavatsky is said to have started a closed esoteric section of the Theosophical Society; Yeats was a member of this group, too.

Olcott's address "My Stand for the Theosophical Society," a classic in Theosophical circles, puts the overall mission of the group squarely: "If I understand the spirit of this Society, it consecrates itself to the intrepid and conscientious study of truth, and binds itself, individually as collectively, to suffer nothing to stand in the way." The Society, a worldwide organization devoted to learning about ageless wisdom, encompasses the study of various religious traditions as well as philosophy, scientific theories and systematic spiritual practice. The three objectives of the Society are paramount to its hundreds of lodges and study groups that meet regularly: "To form a nucleus of the universal brotherhood of humanity, without distinction of race, creed, sex, caste, or color. To encourage the comparative study of religion, philosophy, and science. To investigate unexplained laws of nature and the powers latent in humanity."[18]

Thus we witness in Yeats' writing a real aim to arrive at truths and reveal them to his readers but in symbols that remain somewhat veiled. This recondite approach shouldn't surprise us, however, for Yeats worked hard in metaphysical circles to learn and even to practice magic, so it wouldn't be fitting to give away secrets. He is quoted as having told a friend that next to his poetry, "magic was the most important pursuit of his life."[19] Indeed, Yeats makes his audience work to open sealed messages in his prose; and the unaware may read his essays but miss the mystical connections.

About clairvoyant, mystic Helena Blavatsky, co-founder of the Theosophical Society embraced by Yeats, volumes have been written, but here it is interesting to relate a small story about her with regard to roses. She is said to have had an occasional awful temper; and one time she showed it while in India in the presence of a particularly offensive chauvinist on whose head she made a shower of roses fall.[20]

Returning to Yeats and his significant affiliations, although the Hermetic Order of the Golden Dawn's link to the Rosicrucians is not entirely clear, there is a connection in their overlapping membership. The order has always embraced the rose in all its power and beauty and used it as one of its symbols. Today, less hidden, the Rosicrucians seem to maintain several different groups, one of which has built an impressive network supporting many publications, home study courses, a park in San Jose California, a Rosicrucian Egyptian Museum and even a Rose-Croix University International offering in-depth courses that expand on aspects of Rosicrucian studies, including the universal laws of nature. The university creates opportunities for students' personal development and spiritual growth in a classroom environment instructed by experts in their fields. So, where does Yeats fit in? Like the Rosicrucians and through his awareness of their doctrine, he had more than a cursory understanding of the rose as a symbol; he knew its full meaning and connection to cosmic consciousness. Seward's analysis astutely notes the difference in the tone of Yeats's writings after his initiation.

It's true that not a lot was known about the Rosicrucians during Yeats's time, however, and because several groups exist, there is some confusion about which is which. Initially, to keep its work from becoming adulterated, the organization had a great need for secrecy, but enough information seeped over the transom to assure us that a rose cross symbol was essential in its ceremonies. The initial Rosicrucian manuscripts were said to be first circulated in Germany around 1610; the society was begun "to afford mutual aid and encouragement in working out the great problems of life and in searching out the secrets of Nature; to facilitate the study of the system of Philosophy founded upon the Kabballah and the doctrines of Hermes Trismegistus."[21] An authority on the Rosicrucian Order, Paul Foster Case has explained that the name lives on and even flourishes because people are attracted by what they've heard about the Order and what they think they know of the early manifestoes which are actually very short but not intended to "move gross wits."[22] Often misunderstood, the Order deliberately conceals itself from those whom they fear would be incapable of understanding. Not an organized society like the Freemasons, as Paul Foster Case reminds us, the Order's members do not pay fees, submit applications or participate in a ceremony. According to Case, "The Rosicrucian Order is like the old definition of the city of Boston: it is a state of mind. One becomes a Rosicrucian: one does not join the Rosicrucians."[23] A contemporary booklet published by the organization underscores that this remains the case. The group is said to "encourage open minded questioning and self-mastery" and to promote a higher "way of life."[24]

However we make sense of the foundations of the Rosicrucians or of the Hermetic Order of the Golden Dawn, a rose cross, used as a symbol by both fraternities, most likely provided the essential connection to Yeats's psyche and the metaphysical messages he intended in *The Secret Rose*. First and foremost, the rose has been used as a "sign of silence and secrecy." In his "Brief Study of the Rose Cross Symbol," Thomas D. Worrel explains that "the rose, like the cross, has paradoxical meanings." It is at once a symbol of "purity and passion, heavenly perfection and earthly appetite, virginity and fertility, death and life."[25] What's more, the rose has significance in numerology. According to Worrel, it represents "the number 5 because it has five petals. And the petals on roses are in multiples of five. Geometrically, the rose corresponds with the pentagram and pentagon."[26] In Rosicrucian teachings, the "number 5 represents Spirit and the four elements."

But for what purpose did the Rosicrucians and other related groups like the Golden Dawn exist? Why were they founded in the first place? Lynn Picknett and Clive Prince in *The Templar Revelation* provide an excellent interpretation, showing how the Rosicrucian movement posed a major threat to the Catholic Church. The "idea of mankind's essentially divine status did not accord with the Christian [dogma] of 'original sin'– the idea that all men and women are born sinful because of the fall of Adam and Eve."[27] Thus, the Rosicrucians went underground and became secret out of necessity, not because they wanted to operate in the shadows per se. As a result, although finally noticed in the seventeenth century, the Order had been fully established long before; "in fact it is scarcely an exaggeration to say that Rosicrucianism was the Renaissance."[28]

Amplifying the Rosicrucian's floral preference, Golden Dawn symbols embrace both the lotus and the rose. In consecration ceremonies, adepts wear the Rose Cross Lamen even though the symbol is so sacred to the Order that advice has been given never to touch it with human hands.[29] Intricate and magnificently colorful, this Rosy Cross invests profound meaning in each of its parts, but roses – rings of them– form the center of the cross while each circle signifies the elements, the ancient seven planets, the twelve signs of the zodiac and correspondence to the Hebrew alphabet. So the influence on Yeats's *Secret Rose* is clear.

What's more, a careful study of Yeats's short story "The Secret Rose" makes it almost impossible to miss parallels to the crucifixion of Jesus Christ, or even more recently to the tragic death of well-known spiritual leader Dr. Martin Luther King, Jr. Both died prematurely – as does the protagonist in "The Secret Rose" for "stirring up the people" toward truth, to quote Yeats. Seekers of truth are generally no strangers to dissociation and dissonance in their connection to non-seekers, although it is their aim to connect. The disjunction with others more than likely results from the elevated frequencies that occur when one is seriously on the path of truth and higher awareness.

Published in 1897, Yeats's "The Secret Rose" is an odd, short piece about an outcast bard named Cumhal who was doing what he did best: travel from hamlet to hamlet in a "particolored doublet" singing and reciting poetry. Known in Ireland of the time as a gleeman, young Cumhal described his own soul as "indeed like the wind, and it blows me to and fro, up and down and puts many things into my mind and out of my mind, and therefore I am called the swift, wild horse."[30] Yeats paints Cumhal as an outcast, one who manages late one evening, by maligning the Church, to so offend a certain friar named Coarb that the clergyman decides to kill him by crucifixion. The reason is clear: "the bards and the gleemen are an evil race, ever cursing and ever stirring up the people, and immoral and immoderate in all things, and heathen in their hearts."[31] Obviously, Yeats paints the friar with an inability to see himself: full of mean spirit, hatred and swiftness to judge as he prepares to complete a bleak, horrific act.

The atrocious friar quickly determines that death on the cross is the gleeman's appropriate punishment; Coarb even awakens fellow monks to assist in overseeing the instrument's construction. Friar Coarb wants to spread meanness around; it isn't enough to contain it within himself. But all during the carpentry, Cumhal seems undaunted, not fearful, still a free spirit, for he knows something the friar and his accomplices don't. The gleeman's being is entrusted to the Holy Spirit, as Yeats reveals so subtly in the metaphor of a rose: "... I have been the more alone upon the roads and by the sea, because I heard in my heart the rustling of the rose-bordered dress of her who is ... subtle-hearted and more lovely than a bursting dawn to them that are lost in the darkness."[32] This rose-bordered dress is, metaphorically, the Holy Spirit. Unbeknownst to the friar and his accomplices, Cumhal has already made the intimate connection. Prepared and unafraid to die an earthly death because he knows something profound, Cumhal already has everlasting life.

In selecting character and plot, Yeats confronts ancient, pervasive problems of the Christian Church and the concerns of a number of world religions: deceit, violence, hypocrisy and egoism. The merciless friar condemns the gleeman quickly, so fast in fact that he can't wait until morning for fellow clergy to awaken to have them join him in putting Cumhal to death. In one bold move, the friar makes a 180-degree turn away from the real, true teachings of Jesus Christ of the Christian *Bible*, swifly abandoning the tenets of the faith, – forgiveness, love, charity and tolerance –, to determine that the youth should perish immediately simply because of ill words toward the Church. Yeats sets up his characters provocatively to criticize the clergy and, since the monks acted in collusion, of the Church itself.

In "The Secret Rose," Yeats's selection of adjectives such as "swift, wild horse," pregnant with metaphysical meaning, corroborate this interpretation. The gleeman was not afraid to be crucified because he was safely in a connection with the one and only incorporeal God, what those in the metaphysical tradition would refer to as endlessness or cosmic consciousness. Moreover, Yeats describes the gleeman as having a "bulging wallet," another metaphor indicating fullness with what Christians refer to as "Holy Spirit." And thus the story goes on to a scene leading up to the crucifixion when Cumhal gives away the equivalent of his last supper; the gleeman throws strips of bacon among beggars who fight until the last scrap is eaten. Yeats gives Cumhal few but striking last words to moan in the midst of the grumbling mendicants before he dies. "Outcasts," he calls to their unheeding ears, "have you turned against outcasts?"

From Yeats to Rumi

More than six centuries before Yeats, and nearly a century before Dante used the rose in carnal and spiritual symbolic manifestations, Mevlana Jelaluddin Rumi, an evolved Afghanistan-born spiritual writer, made the same provocative flower an essence in his poetry. Born on September 30, 1207, in Balkh when Afghanistan was still part of the Persian Empire, Rumi was inspired by his father's secret inner life and became a sheikh – a religious scholar who helped the poor.[33] But the pivotal event in his life seems to have been meeting the enigmatic saint Shamsi Tabrizi, his teacher and spiritual guide.

Thanks to a remarkable translation by Coleman Barks, Kabir Helminski and others, we can now experience Rumi's delightful, spirited poetry firsthand, the smooth translation reading almost as if Rumi wrote originally in English. Or is it that Rumi's poetry is so spiritually profound, so inspired, that the intended messages remain clear in translation? For me, Rumi's verses transcend all earthly language with lines so pure they seem palpably connected to Spirit and *intended* to be translated and circulated in a variety of global tongues. Coleman Barks' collection, *The Essential Rumi* beautifully captures Rumi's allegorical and spiritual poetry as it was meant to be. Witness for instance Rumi's "Spring Is Christ." The lines "We walk out to the garden to let the apple meet the peach, to carry messages between rose and jasmine"[34] signify connections between the carnal and spiritual, as do the next several lines that use the rose as a metaphor to reach "a lamp" – the lamp being the "Holy Spirit":

> Spring Is Christ
> Raising martyred plants from their shrouds
> Their mouths open in gratitude, wanting to be kissed.
> The glow of the rose and the tulip means a lamp
> Is inside …[35]

Barks' collection features many roses, but the metaphysical symbolic use of the rose is clearest in Rumi's "Dying Laughing." The poem opens as a lover tells his beloved how much he loves her and how "faithful and self-sacrificing" he is toward her. "There was fire in him. He didn't know where it came from…"[36] But his beloved, not impressed with all the material feats he's accomplished, tells him so, conveying in essence that in outwardly acts he had done very well as a lover, but he hadn't died. To hear that he had not died struck him as funny since he'd done so very much for his intended.

> When he heard that, he lay back on the ground
> laughing, and died. He opened like a rose
> That drops to the ground and died laughing.[37]

Only when the lover dies laughing, a metaphor for letting go of the material world, does he "open[…] like a rose," that is, he makes the quantum leap toward the Holy Spirit. Thus, again and again, Rumi shows us that he, too, held the keys to the ultimate connection long before Dante's epic appeared.

Reminiscent of Cumhal in Yeats's "Secret Rose" or the lover in Rumi who learned to "die laughing," the old man in Hakim Sanai's poem "The Old Man of Basra" struggles constantly to connect to the Holy Spirit. The old man arises daily to the same mundane, unessential questions that millions of human beings worldwide also ask: "What shall I eat?" and "What shall I wear?"[38] The old timer, however, resists such queries whenever they invade his mind. To 'what shall I eat?' he learned to answer "death," and to 'what shall I wear?' he learned to answer "a winding sheet." These responses reveal spiritual evolution; the old man of Basra fought to overcome the material world.

Now, Sanai, writing nearly a hundred years before Rumi, also features roses. This deeply metaphysical poet about whom little is known lived during the reign of Bahramshah (1118-1152) and probably died in 1150.[39] His "garden roses," however, if prominent, are oddly used to juxtapose evil and good; for "self-Cherishers" roses assume the form of "malignant boils." This is like the case for those whose prison consists of "deceit, hatred and envy" – Dante's three beasts. Thus not only Rumi but Sanai, too, preceded Dante and Yeats in knowing precisely what the latter came to know as a result of deep study of the ancient teachings.

In conclusion, having studied allegorical messages in St. Jerome's Latin Vulgate version of the *Holy Bible*, Dante allows Beatrice to mediate his access to the Light, a seemingly ironic choice of guide – she had rebuffed him in real life – that nonetheless suggests the poet's transcendence of mundane travails. Yeats, in his association with the Theosophical Society and the Order of the Golden Dawn, also had teachers or guides to hermetic knowledge, Helena Blavatsky among them. Like Blavatsky's, Yeats's full involvement in metaphysical studies brought him to a high level of understanding about realms far beyond the material world. As the saying goes, when the student is ready, a teacher will appear. For Rumi, Tabrizi arrived when he was ready; for Sanai we can't be sure but it appears that his teachers were Lai-Khur and Yusuf Hamadani.[40] As for Eliot and Joyce, they found guides as well.

And the rose? Although in theory available to all, it remains hard to know. Yet its revelations are needed more than ever given the meanness and lack of humility today. Mystics and adepts were careful not to waste time on "gross wits" and on those who refused to do the necessary work toward what mystically-based secret societies refer to as "self-mastery."[41] Still, human behavior that mimics the rose can trump the material world, as the great writers and poets from Yeats to Rumi knew so well.

References

[1] For example, since the organization was founded in 1928, the Girlfriends, Inc. has used the rose as a symbol.

[2] On coins from the Greek Island of Rhodes in 1200 B.C. we find the rose as a symbol.

[3] For instance, the movie "V is for Vendetta" in which rose petals are used as a powerful symbol of good overcoming evil. Starring Natalie Portman and Hugo Weaving and directed by James McTeigue, "V" was released in 2006.

[4] In Nandor Fodor's, *Between Two Worlds* (West Nyack, NY: Parker Publishing) see Chapter Two "The Riddle of Vaslav Nijinsky" that describes Nijinsky's superior dance skills and levitation in the ballet "Secret Rose." Page 22 describes how "Nijinsky possessed the ability to remain in the air at the highest point of elevation before descending." Although Nijinsky suffered from mental illness later in his life, he is known to have had unusual talent connected to dance feats.

[5] See for instance Umberto Eco's *The Name of the Rose* (NY: Harcourt Grace & Co, 1980), a well-researched work of fiction, and also the movie based on this title starring Sean Connery and Christian Slater.

[6] Barbara Seward, *The Symbolic Rose* (Dallas, TX: Spring Publications, 1954), 37.

[7] Helen M. Luke, *Dark Wood to White Rose: A Study of Meanings in Dante's Divine Comedy* (Pecos, NM: Dove Publications, 1975), 11.

[8] Luke. *Dark Wood to White Rose.* 11.

[9] Dante Alighieri, *"Divine Comedy," Classics Appreciation Society Condensations.* Trans. H. F. Cary (np: Grolier, 1955), 451.

[10] Ibid. 444.

[11] Ibid.

[12] Dante Alighieri, *The Inferno.* Trans. Henry Wadsworth Longfellow. Intro. and Notes Peter Bondanella (NY: Barnes and Noble Classics, 2003), xxxi.

[13] Ibid., xxxv.

[14] Barbara Seward, *The Symbolic Rose*, 156.

[15] Ibid. 89.

[16] Ibid.

[17] Richard Cavendish, *A History of Magic* (NY: Arkana, Division of Penguin Books, 1990), 142.

[18] *Introducing the Theosophical Society*, Wheaton, Illinois. http://www.theosophical.org Retrieved 27 December 2014.

[19] R. Cavendish, *A History of Magic*, 146.

[20] Colin Wilson, *The Occult*, 416-417.

[21] Thomas D. Worrel, *A Brief Study of the Rose Cross Symbol* (np: np, nd), 3. In the author's files.

[22] Paul Foster Case, *The True and Invisible Rosicrucian Order: An Interpretation of the Rosicrucian Allegory and An Explanation of the Ten Rosicrucian Grades.* Boston: Weiser Books, 1985), 4.

[23] Ibid., 5.

[24] *Mastery of Life*, publication of the Supreme Grand Lodge of the Ancient & Mystical Order Rosae Crucis, AMORC, Inc. 7.

[25] Worrel, *A Brief Study ...* 2.

[26] Ibid., 2.

[27] Lynn Picknett and Clive Prince, *The Templar Revelation: Secret Guardians of the True Identity of Christ* (NY: Touchstone, 1997), 134.

[28] Ibid., 135.

[29] Israel Regardie, *The Golden Dawn: A Complete Course in Ceremonial Magic.* Four Volumes in One (St. Paul, MN: Llewellyn Publications, 1986), 310.

[30] James Pethica, ed. "From The Secret Rose, 1897," *Yeats: Poetry, Drama and Prose* (NY: Norton, 2000), 195.

[31] Ibid.

[32] Ibid. 197.

[33] Coleman Barks, et al. trans., "Introduction." *The Essential Rumi* (Edison, NY: Castle Books, 1997).

[34] Ibid. 37.

[35] Ibid. 38.

[36] Ibid. 212.

[37] Ibid.

[38] Hakim Sanai, *The Walled Garden of Truth*, Trans. and Abridged D. L. Pendlebury (London: The Octagon Press, nd), 7.

[39] Ibid., 37.

[40] Ibid., 8.

[41] Hidden knowledge and the secret teachings of Jesus are mentioned and alluded to throughout the *Bible*, including in Mark 4:11 which says "Unto you it is given to know the Kingdom of God: but unto them that are without, all these things are done in parables." That is, many people worldwide read the *Bible* but few fully understand it. A tremendously complex book, it requires years of study, reflection, and pure intent to comprehend. This, too, is made clear in the Book of Mark, Chapter 4, Verse 12: "Seeing they may see and not perceive; and hearing they may hear and not understand; lest at any time they should be converted and their sins should be forgiven them."

13

Rose Vignettes:
Black Plague to Gulag

By Frankie Hutton

In 1938, German-Jewish poet and writer Rudolf Borchardt publish-ed one of his last masterpieces titled *The Passionate Gardner*. In this little known work of about 340 pages, translated into English by Henry Martin, Borchardt uses plants and gardening as a deep ante-cedent for the potentiality of human beings to grow up, to master themselves and to create something good for all humanity. Borchardt beseeches us to understand a "grandiose metaphor in which the human soul is itself a garden, strictly and fervently in the care of God: the heart of the child of the world is rife with weeds, and He plucks them up; the heart of the innocent is a pious modest field full of silent grace, and fragrant with fine perfume…"[1]

It is no coincidence that the *Holy Bible* places a garden among its early scenes. Borchardt reminds readers that Christ's pained depar-ture from the world also took place in the garden of Gethsemane. Thus, "purity can only be found within the protection of vegetation, and everything which is not impure is a garden" where individuals can master and transform their world. Thus, garden and flower metaphors, in Borchardt's view, stand for the great work that human beings can accomplish when, unhindered by weeds, they transcend the ordinary, open up and bloom like so much flora. They find merit in pure work that leads to something good for all human kind.

Constance Casey, reviewing *The Passionate Gardner* for the *New York Times* in 2006, refers to Borchardt's work as a form of "botan-ical globalism" that points us to the heart of an urgent matter in gar-dening: "whether to grow native plants only, or to permit plants from other continents in our gardens." Ultimately growers can decide whether to embrace plants uncommon to their area or settle for pro-vincialism – preferring common rather than exotic plants.

This horticultural metaphor is easily applied to service to huma-nity. Are we to live separately and selfishly for our own families and immediate circles alone, or do we swing outside that realm to create something wonderful and transcendent benefitting all?

Significantly, more than any other flower, the rose lends itself to cross breeding and provides the gifts of beauty, medicine, fragrance and perfection. In this sense, the flower serves humanity both aesthetically and pragmatically. Available to everyone, it suggests we can choose to exit the familiar and connect with, help, and support what may seem foreign. Reaching beyond themselves, a few admirable people have understood Borchardt's botanical globalism and, often through adversity, created good, lovely and useful works. Models of selflessness, they make an intense effort to serve others for no material gain. Granted, this is difficult work whose importance, sadly, some of us fail to understand.

Borchardt's horticultural metaphors, however, allow us to find those few paragons whose embrace of the rose includes healing, music, social justice and world peace, for the rose is a singular symbolic presence in these. In metaspiritual matters as well, the rose is a useful symbol of the opening of the chakras, those urgent, unseen energy centers in the human body that average human beings never come to know.

When we consider the rose's vastness and universality, the flower no doubt reaches its earthly apex in the work of Daniel Andreev who sees it embracing all people in one deep, moral force for world peace. Andreev and others like him reveal their knowledge of the flower's qualities. Engendered by a strong spiritual connection, their acute awareness often followed revelations manifest in light that vibrantly revealed itself when they were ready.

Found in diverse venues and nearly all epochs, extraordinary people who have understood the urgent need for lives of service have made their legendary marks in part by clasping the rose, not only symbolically, but also practically since the botanical essence has been found to possess magnificent medicinal properties.

Dr. Edward Bach
A British physician who studied medicine at University College Hospital, London, Dr. Edward Bach (1886-1936) understood the medicinal rose. After abandoning his traditional, successful practice, he found plants as a curative and persisted in his exploration even though his work with plants was antithetical to the traditional

Western approach to medicine in which he had been trained. In fact, he is said to have come to a psychic understanding of plant essences. The flowers he used are thought to be of a "higher order" since they contain frequencies within the human energy field.[2] This is perhaps a novel idea for readers who have never considered flowers more than lovely to look at, smell or use to adorn dining tables and living rooms. Yet, the human soul is thought to contain 38 qualities or virtues identical to qualities of flowers selected by or "given to" Dr. Bach by a higher realm. The energy or divine sparks associated with this flora are like wavelengths in the energy field that Bach came to know intuitively.[3]

Hence, having determined this likeness, Dr. Bach introduced the essence of wildflower blooms, including rock rose and wild rose, to treat all sorts of ailments, including asthma, irritable bowel syndrome, headaches, muscular tension, rashes and more. Bach flower essences, a spin-off of the ideas of homoeopathy but said to be more pure because they lack the products of the disease, are now carried in top health food stores worldwide and are available through mail order. The essences can also be used to treat mental upsets such as remorse or lack of confidence. With these 38 powerful flower essences in mind, moreover, Dr. Bach based his work, akin to herbal medicine, on awareness that "disease is in essence the result of conflict between Soul and Mind."[4] Yes, Bach admits, bacteria play a part in the spread of physical disease, but they are only part of the equation because not everyone exposed to these agents gets sick or develops disease; factors higher than the material world must be considered.[5] In his essay "Heal Thyself," Bach explains that disease reveals our faults so that, once we can correct them, the malady will have been beneficial and open to relief:

> No effort directed to the body alone can do more than superficially repair damage, and in this there is no cure, since the cause is still operative and may at any moment again demonstrate its presence in another form. In fact, in many cases apparent recovery is harmful, since it hides from the patient the true cause of his trouble, and in the satisfaction of apparently renewed health the real factor, being unnoticed, may gain strength.[6]

Throughout recorded history, a few extraordinary people like Edward Bach have mastered themselves and the feat of connecting with others for the purpose of advancing work that aids humanity and, as we are beginning to see, so many of these individuals from countries and cultures far and wide have in some way used or embraced the rose.

Nostradamus, 16th Century France

Although surrounded by detractors and disbelievers, the great Michael de Nostradamus led an extraordinary life, and his powers seemed to increase as the years progressed. From the universe, Nostradamus was given the unusual, bitter-sweet gift of prophecy and, in the past five hundred years, neither his productivity nor predictions have gone unnoticed. A number of films, books and articles have chronicled his life and work. More than an astute physician, clairvoyant and mage, the Frenchman, known simply as Nostradamus, predicted the explosion at the World Trade Center in New York and numerous other devastating events over the past centuries. Destined to be as controversial as he was sought after by the rich and royal for his psychic powers, he was faulted by his wife's kin for inability to save his own family from the Black plague that struck Agen in southern France in 1537. For this failure, he was asked to return the dowry his wife's relatives had paid when they married. An Italian doctor and apparently jealous friend, Julius-Cesar Scaliger (1484 - 1580) provoked a bitter quarrel between Nostradamus and his in-laws which shattered their friendship; this hurtful time was followed by a charge of heresy and exile from Agen.[7]

Ian Wilson's fine biography of Nostradamus provides an excellent translation from the treasurer of Aix-en-Provence when the plague hit that town in 1546, less than a decade after it struck Nostradamus's family. Victims suffered greatly from "Black egg-sized swellings or buboes would appear in the armpits and groin, oozing blood and pus. Black blotches would follow all over the skin accompanied by severe pain and within five days the victim would usually be dead."[8]

Although there was said to be no known cause nor cure, conditions ushering in the plague were a combination of strange weather patterns and unsanitary living habits that left Europe ripe for an epidemic. Peter Lemesurier, a Cambridge-trained linguist and teacher, explains how the plague's prelude, "Europe's Little Ice Age was in full swing. As a result winters were tending to become more arctic, summers colder and wetter. Even when plowing and sowing could be done at all, the crops would often rot in the fields. No crops meant no food, and no food meant no survival."[9] This unusual weather caused France's agricultural cataclysm as the territory suffered "no fewer than thirteen major famines during the course of the century."

But as a result of unaccustomed climactic patterns Europe also endured flooding, and the over-abundance of water coupled with trauma to the agricultural system spelled disaster. By 1544, heavy rains caused the Rhone River to rise, thereby inundating towns and cemeteries; corpses washed out of their graves, and exceptionally awful sanitation problems also encouraged the plague.[10]

Although Nostradamus' predictions and his life's story have been well recorded, information is scant concerning his clever mixture of rose petals as a preventative during the plague years. Nonetheless, as Ian Wilson narrates, Nostradamus authored a *Treatise on cosmetics and jams* that prescribes use of the rose as a prophylaxis for those who had not yet fallen ill.

> Take some sawdust or shavings of cypress-wood, as green as you can find, one ounce; iris of Florence, six ounces; cloves, three ounces; sweet calamus [cane palm], three drams; aloes-wood six drams. Grind everything to powder and take care to keep it all airtight. Next take some furled red roses, three or four hundred, clean, fresh and culled before dewfall. Crush them to powder in a marble mortar, wooden pestle.[11]

Nostradamus then advised adding more half-unfurled roses to the mixture and rose juice as well. Once made into a pill, it would smell good, too. What the great mage did not explain and may not have known in the scientific-medical language of his time is that rose petals contain important phytochemicals such as beta-carotene, botulin, catechin, flavonoids and nutrients such as calcium, iron, magnesium, manganese, phosphorus and potassium; the petals also offer an array of B vitamins and are generally good against all kinds of infections.

Dr. Tomin Harada, World War II Era, Japan

Few outside of the medical profession in America or Europe have heard of Tomin Harada, for before World War II he was just another professional serving his country as a military physician. The World War II era changed all that. Born in 1912, Harada graduated in 1936 from Jikei Medical School in Tokyo and soon after, in 1938, was thrust into war with China, one of the opening stages of World War II. As a military officer-surgeon, Dr. Harada put himself in a tense situation when he was asked to talk to a group of soldiers using a canned model prepared for him for such occasions. "... I just could not bring myself to pass on such nonsense."[12] So deciding to speak from the heart, Dr. Harada remembers his speech going something like this:

> Although the Japanese and Chinese races are slightly different, we, along with the Tungas people, the Koreans, and the Melanesians, all have Chinese blood flowing in our veins. More importantly, Japan has received much from Chinese culture. We are now fighting, but we must not let this become a war of hatred. The time will surely come when we will be friends again. We are now going to help create an opportunity for that friendship to be reawakened.[13]

In light of his officer's training, Harada was later embarrassed by the speech but had not been able to state what he didn't believe. He had been critical of Japan's notion that it should rule the world just as he resented the colonialism of Great Britain, the United States and Holland. Stationed at the Taiwan Henchun garrison amid poisonous snakes and mountain leeches, with few remaining rations and emaciated from the long campaign, Harada remembered the day the news came that the war was over. He was also shocked to hear that Hiroshima and Nagasaki had suffered Atomic bomb attacks. Eventually returning to his hometown, Hiroshima, he found it obliterated and his family missing. With the aid of a friend, Harada eventually found his kin at their ancestral home outside of town and soon after, in 1946, he built a hospital to help survivors of the devastating blast.

Only after the war and the death of his wife did Harada became a passionate peace activist and, in retirement, having met another former WWII officer in the UK who introduced him to rose cultivation, promoted growing roses in connection with peace activism. Harada felt an instant affection for the lovely, fragrant flower and became a breeder; both men agreed that roses could be used as a symbol of world peace. It is noteworthy that two former officers who had seen the horrors of war from opposing sides could now embrace the simple, fragrant rose as a symbol of global harmony.

Dr. Harada's superior reconstructive, plastic surgery to alleviate the suffering of Japan's victims of the bomb and later his treatment of Vietnam War maladies caused by Agent Orange and napalm were progressive and legendary. He is said to have been especially attracted to roses because the fresh flower reminded him of the young and beautiful children who had been killed in the war and would never have the opportunity to lead the world toward peace. In cross-breeding roses, Dr. Harada developed new brands to which he gave unusual names such as "Phoenix Hiroshima, Hiroshima Mind, Miss Hiroshima, Peace Maker, Red Hiroshima, Dr. Tomin, May Peace, Spirit Hiroshima…" that still exist in Hiroshima and in select gardens all over Japan.[14] In fact, so committed was Dr. Harada to the peace mission that he sent flowers he had bred to world leaders with the wish for no more nuclear war and a "free" world. "Let the roses speak for themselves," he explained. "Roses embody peace and beauty by themselves." His rose initiative caught on in Japan, and the peace movement in that country today owes a debt to his untiring work with roses for peace. He died in June 1999.

Mozart (1756-1791), Austria

Although the rose is a powerful symbolic presence in the universe, its semiotic connections are easily missed by the profane and noisy whose high vibration music blocks them out, probably because the flower itself is quiet and refined. Yet in exquisite compositions, song writers and composers throughout the world have embraced the rose, and we can deduce from a couple of examples that at least some of the flower's adepts were no strangers to higher awareness.

This is the case with Wolfgang Amadeus Mozart. Careful study of his life and work reveal an intimacy with mysticism, and colleagues may have resented his displaying what they felt should remain under wraps. Speculation continues about Mozart's breaking silence concerning secrets revealed to him in Masonic initiation, an 'indiscretion' that may have led to his mysterious, untimely death.

A child prodigy, Mozart, who mastered the violin, harpsichord and organ, went on to compose some of the greatest classical music of all times. The evidence is that during his short life, he knew the mystic rose. One of his greatest renditions, finished between 1790 - 91 shortly before his death, *Die Zauberflöte* or *Magic Flute* speaks volumes about the rose's arcane signals. Although unnoticed by most opera-goers, the opus conveys important Masonic and mystical messages that naturally include the rose.

Important tenets of Freemasonry – service, silence, patience, steadfastness – appear in *The Magic Flute*. It is not by chance that Judy Taymor's breathtaking production at New York City's Metropolitan Opera during the 2007-2008 seasons prominently depicts a rose garden in act two, scene three. Taymor, an award winning, renowned theatrical producer, was so astute as to make the roses lighted, crystal, red and colossal. The rose symbolizes love, discretion, and the potential for connection with the Source of all and everything; it is truly the mystic rose. Both the garden and the rose are potent symbols that can lead to the light of Cosmic Consciousness – for those blinded by deceit, ego, and greed to becoming one with the Source, a truth deliberately obscured by denominational divides created to manage, placate, and manipulate most people.

In *The Magic Flute*, lovers Pamina and Tamino overcome evil in their midst and grow tremendously in their understanding that true love is unconditional and never too possessive. These lessons challenge Pamina who can't comprehend Tramino's silence during his initiation to higher consciousness. She lacks the discipline he displays in staying the course to become a master, first and foremost, of himself. Protected by the magic flute and aided by three spirits, Pamina eventually finds Tamino to accompany him through many ordeals connected with water and fire. And before the final curtain, they wed, but this consecrated union, a mystical symbol, is often misunderstood.

Now, Pamina and Tamino's union teaches us much about the meaning of marriage. For, in Hodson's words, "Marriage and intercourse, whether legal or illicit, [can] refer not to any carnal relationship but to a spiritual 'marriage', or blending of consciousness, at any level."[15] Thus, people's potential to rise up and fuse with the light, to transcend darkness, is suggested by human unions, the goal being to fuse with Cosmic Consciousness, making solo souls more than they would have been alone, even if this melding rarely happens on earth. Without schooling in the institution's true significance, we need to learn that the aim of marriage is to *symbolize* the sacred. To succeed at matrimony, both human partners must invest focused effort, in the same way intense work prefaces accession to the Holy Spirit's light. Yet despite metaphors and symbols like the rose and the cross that could tutor humanity to 'get it', the troubled practice continues worldwide. Alas, instead of working, people often prefer talking; they intellectualize the scripture, allowing their egos to get in the way of harmonious union with another human being and hence with the Cosmic realm.

Given the intense commitment demanded, most people who would rise find their discipline falls short. Likewise marriage, including sex, symbolizes more than itself. Inherent in earthly couplings is a larger truth: that connection to one means unity with many, and the love and support in the tiniest unit echoes aspiration toward the Source of all. To war with each other is not why we are here, a truth that may remain elusive given the vast effort to withhold it from us.

Charles Austin Miles, New Jersey, USA

At this remove, little is known about American Charles Austin Miles (1868-1946). Most people have never even heard of him, but practically all in the Protestant Christian tradition know his song "In the Garden" as a staple of Sunday church services. In 1912, the lyrics and tune were "given" to Miles who had been asked to write a hymn "sympathetic in tone, breathing tenderness in every line; one that would bring hope to the hopeless, rest for the weary, and downy pillows to dying beds."[16] Guided by a vision of Jesus when he opened his Bible to John 20, Miles found the meeting of Jesus and Mary. Seated in a dark room at the time, the lyricist reports on composing the famous song:

> I awakened in full light, gripping the Bible, with muscles tense and nerves vibrating. Under the inspiration of this vision I wrote as quickly as the words could be formed the poem exactly as [it] has since appeared. That same evening I wrote the music (in *101 Hymn Stories*).[17]

Like the writer William Butler Yeats, and according to his 1946 *New York Times* obituary, Charles Austin Miles had Masonic connections.[18] Born in Lakehurst, New Jersey, Miles authored 3,000 hymns, among them "In the Garden" which was not a financial success in his lifetime. It has since sold three million copies, however, and recordings exceed one million. The hymn has worldwide distribution and is sung in many languages.[19] Because "In the Garden" Miles is precise in its lyrics embracing roses in a garden, we have grounds to speculate that its author had joined a secret society such as the Rosicrucians or the Order of the Golden Dawn since both fraternities fully embrace the rose and, like Miles, would have understood the dew on the hymn-writer's roses as symbolic of divine light. Thus, Miles knows something about Cosmic Consciousness and, given his description of how the song was "given" to him, had tapped into it.

> I come to the garden alone
> While the dew is still on the roses
> And the voice I hear falling on my ear
> The song of God discloses

In the Bible, a garden symbolizes paradise, and the fully-opened red rose stands for connection of the key crown charka to its synonyms, cosmic consciousness, the universal mind, Christ consciousness or God. And as the author shows that he knew, we must "come to the garden alone."

Daniel Andreev

In his realm – the mid-twentieth century metaphysical community in Soviet Russia –, Daniel Andreev masterminded an incredible work known simply as *The Rose of the World*. For quite some time, no one in the West had heard of it nor had more than very few in Russia, because by 1947 Andreev had been sentenced to 25 years in a gulag for outspoken anti-Soviet ideas. Although Andreev's pre-gulag work was destroyed, his opus *The Rose of the World*, written in prison and hidden by friends, survived. Buried until the collapse of the Soviet system, the book appeared first in Russian in Moscow and now in English.

And no wonder the harsh Soviet regime repressed Andreev and censored him. In *The Rose of the World*, Andreev minces no words about the evils of power-hungry government officials. He gets to the heart of heady political and economic matters that beg to be confronted by citizens everywhere, now more than ever. In Jordan Roberts' recent English translation, Andreev holds that human beings should be tremendously concerned by the imminent formation of a global police state. Writing sometime between 1947 and 1957, the author saw that new technological advances simply furthered this process:

> It should come as no surprise today that one side of every scientific and technical advance goes against the genuine interests of humanity. The internal combustion engine, radio, aviation, atomic energy – they all strike the bare flesh of the world's people with one end, while advances in communications and technology enable police states to establish surveillance over the private life and thoughts of each person thus laying an iron foundation for life-sucking dictatorial states.[20]

It is eerie to recall Andreev's warnings now, almost fifty years later, considering American writer Norman Mailer's denunciation of technology. Mailer, too, excoriated a mechanized world where "souls are increasingly interchangeable."[21] Whether true or not, we cannot know at this remove, but technology often hurts as much as it helps, like the Southern story of a cow who, having given a splendid bucket of milk, takes her hind leg and kicks it over. Simply consider the tyranny of cell phones, iPods, and computers that obscure their tracking functions and invade our privacy. The breach of quiet caused by cell phone use is inconsiderate; the spread of these devices may be out of control. Coupled with high tech weapons, so-called "technology" becomes increasingly detrimental once radiation is factored in. Thus Andreev's admonishments become all the more chilling, especially in light of recent televised Congressional hearings investigating a new mercenary security force called Blackwater, USA. A super-charged, highly trained group with the latest high tech weapons and know-how, Blackwater, at the government's bidding, will go anywhere and do anything.[22]

Released early from prison after a decade of poor diet, beatings and cruel treatment that left him broken and ill, an emaciated Andreev knew that his days were numbered, yet he persevered to finish *The Rose of the World*. Translator Mikhail Epstein, who has interpreted a large portion of Andreev's work, reminds us that the author's preoccupation with nature drives much of the philosophical under-pinning of his oeuvre. As Epstein concludes:

> Throughout his creative years, Daniel Andreev suffered under the pressure of official Soviet ideology's "stubborn iron materialism," but his inner resistance to this mysticism of *materiia* did not push him to the other extreme of bodiless spiritualism. Nature was the center of his whole system, and he singled out a special category of "elementals" *(stikhiali)*, spiritual entities that have an elevating effect on the human soul and are embodied in such natural elements *(stikhii)* as rivers, trees, wind, and snow. Daniel Andreev enjoyed travel-ing through the wildest and most remote Russian forests, because for him nature suggested the most genuine way of knowing God...[23]

Andreev's marriage survived the gulag, but he didn't live very long after release. Incarcerated in the same system, Alla Andreeva had suffered, too, although at different locations away from her husband. Ann Applebaum's award winning *Gulag* provides insight into the horrors of the Soviet prison system. Alla Andreeva remembers her time in the gulag as demeaning and harsh. Indeed, gulag imprison-ment was engineered to strip every ounce of individuality and dignity. For instance, Alla Andreev's first camp allowed prisoners to wear their own clothes, but beginning in 1948 she was forced to don a smock that she found offensive.[24]

To avoid a global government and police force that he foresaw, Andreev advocated a strong moral body. In *The Rose of the World* he appears to plead that

> We ... recognize the absolute necessity of the one and only path: The establishment, over a global federation of states, of an unsullied, incorruptible, highly respected body, a moral body standing outside of and above the state. For the state is, by its very nature, amoral.[25]

The Global Anti-FGM Movement
To this point we have seen the rose's positive potency, but Pulitzer Prize-winning American author Alice Walker, among others, knew it

as both positive and negative, lovely and fearful, that is, as a universal symbol for the vulva.

Its literary history clarifies this association, as Tobe Levin shows in Chapter 2 of this volume. As Levin points out, the connotations are clear, for instance, in Romantic Poet William Blakes' "The Sick Rose": "O Rose thou art sick./ The invisible worm,/ That flies in the night/ In the howling storm:/ Has found out thy bed/ Of crimson joy:/ And his dark secret love /Does thy life destroy."[26] European history demonizing women partakes of a tradition that also vilifies female genitalia, as Blake does here, depicting sexual desire as lethal.

Alice Walker saw this, too, but set out to correct it, dedicating the first novel by a famous author denouncing FGM to the "blameless vulva." Published in 1992, *Possessing the Secret of Joy* features an insane survivor of infibulation who attracted the attention of activist Efua Dorkenoo, OBE (1949-2014). Dorkenoo contacted Walker. "I wrote to Alice," she explained to Levin, "[because] FORWARD [Dorkenoo's NGO] counsels women like Tashi whose mental anguish at having suffered mutilation has become unbearable. Tashi is so real that I wanted to let Alice know ..."[27] A June 1995 board meeting of FORWARD praised Walker's influence on the struggle to eradicate the practice.[28]

Regarding the rose, Dorkenoo's book *Cutting the Rose: Female Genital Mutilation, the Practice and Its Prevention* recognizes the flower as a privileged symbol in the developing global abolition movement. Artists, too, like Nigerian Godfrey Williams-Okorodus, highlight the rose in paintings against FGM. Okorodus, together with Joy Keshi Walker, spearheads an international traveling exhibition of watercolor, oils and sculpture addressing this grave health hazard and human rights abuse.[29] "Art against FGM. International protest against Female Genital mutilation" (originally "Gezeichnet von der Tradition") similarly deploys visuals in a project launched by illustrators who banded together to protest; in several pictures the rose can be seen.[30]

The rose, in fact, represents any number of anti-FGM movements, and Amnesty International Ireland's 2007 campaign is among the most salient. "... A series of three posters for Amnesty International's [fight] against female genital mutilation (FGM) ... show[s] a different colored rose, with different stitching, [to illustrate] the variations of FGM across the world."[31] Emily Winter interprets these:

... The logo from the END FGM European Campaign by Amnesty International [images] a stitched-up rose. [One] rose is red, symbolising blood [but] the paler white rose [is] better as a white rose could [stand for] innocence, and the mutilation of innocence is the primary psychological consequence of female genital mutilation.[32]

Viewers will lose their virginity as well when, in only 38 seconds, PLAN UK shows the rose, poignant and shocking, in its battle against the scourge. A full, fat flower appears on the screen; scissors, in a single snip, amputate its petals.[33]

To conclude, if roses mean enlightenment, harmony, and peace, then aggression in the name of the rose must end, and in this project, too, as in 'living the rose', everyone has a part.

Coda

No other flower has the unusual talent enjoyed by the rose: to facilitate humans' connection to the Source. Spanning medicine and the humanities, its gifts embrace creative genius that produces beautiful and timeless music, opera and script. Humanity intended to shadow the rose can fully do so when serving others or creating something lovely and beneficial for the good of all.

With the author's appreciation for assistance and editorial support from Tobe Levin (Germany), Hisae Ogawa (Japan) and Alex Prodovikov (Russia).

References

[1] Rudolf Borchardt, *The Passionate Gardener*, Trans. Henry Martin (Kingston, NY: McPherson, 2006), 8.

[2] Mechthild Scheffer, *Bach Flower Therapy* (Rochester, VT: Healing Arts Press, 1988), 16-17.

[3] Ibid., 16.

[4] Edward Bach and F. J. Wheeler, *Bach Flower Remedies* (New Canaan, CT: Keats, 1997), 7.

[5] Ibid.

[6] Ibid., 10.

[7] John Hogue, *Nostradamus: New Revelations* (Rockport, MA: Element, 1994), 16-17.

[8] Ian Wilson, *Nostradamus: The Man behind the Prophecies* (NY: St. Martin's Griffin, nd), 50.

[9] Peter Lemesurier, *The Unknown Nostradamus* (np: John Hunt Publishing Ltd., 2003), 62.

[10] Ian Wilson, ibid., 45.

[11] Ibid., 47.

[12] Tomin Harada, *Hiroshima Surgeon* (Newton, KS: Faith and Life Press, 1983), 4.

[13] Ibid.

[14] Facts provided by Hisae Ogawa, trans. Dr. Harada, *Moment of Peace* (in Japanese) (np: Gariver Products Co, Ltd., nd); Dr. Harada, *Hiroshima Roses*. Trans. Hisae Ogawa (np: Mirai-sha Publisher, 1989); and Dr. Harada, *Chasing the Dream of Peace*. Trans. Hisae Ogawa (np: Kei-shobo Publisher, 1983).

[15] Geoffrey Hodson, *Hidden Wisdom in the Holy Bible*, Vol. I (Wheaton, IL: The Theosophical Publishing House, 1963) 129.

[16] Kenneth W. Osbeck, *101 Hymn Stories* (Grand Rapids, MI: np, 1982), 124.

[17] Ibid.

[18] *New York Times*, March 12, 1946, 26.

[19] Ibid.

[20] Daniel Andreev, *The Rose of the World*. Trans. Jordan Roberts (Hudson, NY: Lindisfarne Books, 1997), 10.

[21] Read Michael Lennon's interview with Mailer in "The Rise of Materialism," *New York Magazine*, October 15, 2007, 24-29.

[22] Of interest is Jeremy Scahill, *Blackwater: The Rise of the World's Most Powerful Mercenary Army* (NY: Avalon Books, 2007).

[23] Bernice Glatzer Rosenthal, *The Occult in Russian and Soviet Culture* (Ithaca, NY: Cornell UP, 1997), 336.

[24] Ann Applebaum, *Gulag* (NY: Penguin, 2006), 177.

[25] D. Andreev, *The Rose of the World*, 10.

[26] William Blake. "The Sick Rose." Web. http://www.poetryfoundation.org/poem/172938 Retrieved 7 January 2015.

[27] Tobe Levin Freifrau von Gleichen, "Alice Walker: Matron of FORWARD." *Black Imagination and the Middle Passage*. Eds. Maria Diedrich, Henry Louis Gates, Jr., and Carl Pedersen. (NY: Oxford UP, 1999. 240-254), 241. For more on Walker and FGM, see Tobe Levin, ed. *Waging Empathy. Alice Walker,* Possessing the Secret of Joy *and the Global Movement to Ban FGM*. (Frankfurt am Main: UnCUT/VOICES Press, 2014). See also Tobe Levin and Augustine H. Asaah, eds. *Empathy and Rage: Female Genital Mutilation in African Literature*. (Canterbury, England; Ayebia Publisher, 2009) for an account of Levin's initial outrage upon learning of the practice.

[28] Pratibha Parmar and Alice Walker also made a film *Warrior Marks* (1993) guided by African activists of Efua Dorkenoo's stature, but both book and movie received considerable flack, including death threats against Dorkenoo and Walker.

[29] Tobe Levin, *Through the Eyes of Nigerian Artists: Confronting Female Genital Mutilation* Exhibition Catalogue. (Frankfurt am Main: FORWARD-Germany, 2006).

[30] "Art against FGM." Web. http://www.art-against-fgm.com/start.htm Retrieved 7 January 2015.

[31] Faith. "Photography: End Female Genital Mutilation." Visualizing Women's Rights

in the Arab World. Blog. https://visualrights.tacticaltech.org/content/photography-
end-female-genital-mutilation Retrieved 7 January 2015.

[32] Emily Winter, "Female Genital Mutilation – Mutilated Innocence." Web.
https://pub209healthcultureandsociety.wikispaces.com/Female+Genital+Mutila-
tion+-+Mutilated+Innocence Retrieved 7 January 2015.

[33] Plan UK. "The FGM rose: #FGMrose - YouTube."
www.youtube.com/watch?v=v3NdP1oj0Y0 Retrieved 7 January 2015.

14

"...Blossom As the Rose"

in *OAHSPE*, *The Emerald Tablets*, and the *Holy Bible*

By Frankie Hutton

The wilderness and the solitary place shall be Glad for them; and the desert shall rejoice, and Blossom as the rose.

Holy Bible,
Saint James Version, Isaiah 35, Verse 1

In obedience to the law, the word of the Master Grew into flower.
The Emerald Tablets

The wilderness and the solitary place shall be Glad for them; and the desert shall rejoice, and Blossom as the rose.

Book of Saphah,
Oahspe

According to an expert on cultural icons and symbolism, the rose holds a "royal status among flowers" explained by "its association with comfort, generosity and discretion."[1] But it is apparently not well known that high level entities in nature and in the universe have been connected to the rose. Inextricable, for instance, is the connection of the rose to the sun, to the great mystic Jesus Christ who was known as the "rose child" and to the floral symbol of the female vulva.[2] Even the magnificent scent of the flower has been revered by both pagans and patricians over epochs and across global timelines. Aware of the importance of aroma, Donald Tyson's historical compilation of occult philosophy reminds readers that in every good matter such as love and good will, "there must be good fume, odoriferous and precious" and, likewise, wherever anything bad or at least of no good value is brewing, there are "stinking fumes that are of no worth. "During ancient times, the essence of the rose was mixed with all sorts of concoctions including "musk, red coral, ambergris" and even with unmentionable animal parts to create "suffumigation," according to the foundations of western occultism.[3] In modern times, the precious oil essence of rose petals has been mixed with numerous other scents to create sought-after fragrances that have brought fortunes to the global perfume industry.

While the rose has been well recognized for its ostensible natural qualities and has been used as a diverse symbol globally, mystical aspects of the flower have been the least known and apparently, for most, unknowable. Nevertheless, a bevy of hints about arcane dimensions have been given to humanity such as the rose's enticing scent and, more than any other flower, its readiness to be crossbred. Another clue lies in the petals themselves: smooth, fragrant and perfect.

A most unusual sign of the rose's special mystical place in the universe is the fabulously colorful rose nebula. This giant cosmic phenomenon spans 50 light years and lies hovering in the cosmos 4,500 light years away from earth, near the constellation Monoceros.[4] Yet few humans have taken note of its presence, probably because it can be seen only with the aid of a powerful telescope. Of the myriad shapes this provocative nebula might have taken, the rose became the resonant form in both magnificent red color and shape. It has been suggested that the nebula's presence was foretold eons ago but no one knows for sure just how long this giant collection of star dust has been in the cosmos. The point of bringing the nebula to mind is that the rose symbolically and botanically reappears throughout the universe in a variety of ways to render its message for those who have "eyes to see." That is, individuals who have arrived at higher awareness perceive the rose's remarkable presence; one of the most important, natural, archetypical symbols ever rendered, the bloom has attracted observers for millenia.

Thus, appearances of the rose or its likeness in the cosmos and in varied sacred literatures bespeak the flower's pedagogic role. It is truly the flower of perfection and love, an apropos symbol for the global village. The mission of the rose transcends any particular religion or belief system and over time has been connected to Isis, Aphrodite, Venus, Mary, the Knights Templar, the Rosicrucians, the Order of the Golden Dawn and even to Zen.[5] Alchemists are said to have embraced it as "the flower of knowing" or "rosarium philosophorum." Its message is irrespective of religion; those who are mired in the dogma of any political or rigidly codified holy persuasion are the most unlikely to understand or to follow the real messages symbolized in the flower. In other words, those steeped in fundamentalist doctrinal matters so fully as to think and live separately and clannishly are blinded to the meanings. The rose's missives cannot reach the zealots of any religion, nor are they for the avaricious. Simply put, the rose means too much for the fanatic, provincial or greedy to sense. Yet the flower outstrips in frequency, as rendered by nature, any particular religion or dogma; its communication transcends particular belief systems and cultures – as contributors to this collection have made clear.

This last chapter in the anthology is in no way the final word on the rose, for there is much more to reflect upon and to understand as each individual who comes, in time, to know the flower intimately will see. The evidence is that striving for higher awareness can open the door to knowledge of all that the rose symbolizes as it ultimately guides toward intensifying consciousness – no small feat!

As we have seen, a number of extraordinary human beings have come to know the arcane life of the rose and embraced the impeccable flower. To amplify the previous chapter about these world class individuals, this final essay focuses on the rose's presence in important sacred literature. Here we renew acquaintance with high level human beings connected in some way to the rose and, in doing so, pose certain questions: why are these people all unusually talented leaders, larger than life, so to say? Is that by design? Is it mere happenstance? Is there some organization or association that unites these individuals? Apparently some relatively few human beings have been pulled toward the flower and all that it signifies, in some cases after years of odd, seemingly serendipitous experiences and occurrences that have ushered them in its direction when the time was right. Sometimes initiation to an esoteric society or fraternal group provided their introduction to the rose; sometimes not. One thing, however, is certain; the flower appears after hard work and honestly-intended truth-seeking begin.

Knowledge and understanding of the inconspicuous presence of the flower in sacred and esoteric literature coupled with vignettes of little-known humans who've encountered it permit insight into the rose's cosmic, mystical realms. But deeper knowledge of the flower emerges only from serious, focused and persevering work on oneself, this singular fact having been conveyed in numerous places, the message both veiled and unveiled. Its hidden sacred knowledge most assuredly not intended for the profane, the rose has thorns which in turn suggest that in part the flower signifies certain limitations, such as an inability to see oneself, a trait of those who also lack regard for others and some measure of discipline in every aspect of their lives.

However, for those 'with eyes to see', the rose drapes its quiet presence all over the universe, in important sometimes veiled literary passages and scriptures, the flower's overt presence particularly striking in *OAHSPE*, a unique and little known nineteenth century Bible.

OAHSPE is mentioned in *The Secret Doctrine*, Helena Blavatsky's two-volume opus that ushered in the late nineteenth century theosophical movement and is alluded to in *The Emerald Tablets* said to be an otherworldly document that pre-dates Christianity. The rose even appears twice in some versions of the Christian *Holy Bible*. Yes, the evidence is that this quietly perfect flower has a varied presence all over the universe, in what appear to be spiritually inspired documents serving diverse cultures as far back as the lost colony of Atlantis thousands of years before the birth of Christ. The flower's veiled presence in some versions of the Christian *Holy Bible* is remarkable too, mostly because the verses containing the rose metaphor in "Song of Songs" and "Isaiah" are generally misinterpreted and apparently misunderstood.[6]

Now, granted, most cultures know at least something about the red rose – its loveliness to touch, to smell, to enjoy and even to eat –, yet the masses remain oblivious to the ubiquitous flower's veiled revelations and to oblique, quiet references to it in important, urgent literature from various epochs and religions. Inimitable and timeless, the rose has a cosmic connection as no other earthly flower, except perhaps its counterpart, the lotus of the East, said to have a thousand petals and at times referred to as the rose-lotus. The two flowers appear synonymous in metaphysics and higher consciousness, although some groups such as the Order of the Golden Dawn use both the lotus and the rose in different ways during ceremonies.[7] But to return to the main point, what are the rose's messages and for whom are the revelations intended?

The rose has been "planted" to offer humans something providential and essential from the universe -- from Cosmic Consciousness. But clarity emerges only once commercial qualities of the flower have been put aside and this essential floral essence is seen as humanity's servant, offering the invaluable gift of natural beauty and quiet perfection – precisely the qualities human beings were intended to provide for each other and the universe. Although bequeathed so many types of roses, from the smallest, most simple wild rose appreciated by Native Americans to the giant rose nebula thousands of miles away, most humans have neglected to create on earth the good that the rose symbolizes in the cosmos.

Yet, symbolically, the rose links to divine aspects of the universe – a sort of quantum theology in a flower. A select few in every generation have understood the splendid messages and tried to live and teach them to those who seem ready or astute enough to understand but often fall short. Social philosopher Joseph Needleman, for instance, in his new book *Why Can't We Be Good?* addresses something of this phenomenon -- of never quite getting there, never quite making the connection to higher consciousness: "The obligation that is offered to us is to strive with all our being to serve what is good – while at the same time, also with all our being, to suffer in full consciousness the naked fact that it is beyond our strength. Then and only then can moral power be given to us."[8] As I hope you will come to see, the rose's symbolic messages are profound and concerned as much with matters of culture, world peace and inner harmony as with individual aspiration to connect with the Divine –, interconnected goals, as highly evolved people come to understand. Connecting the dots, you can glean this through a higher form of intuition that transcends the material world. Insights begin to come slowly after sincere work in the "right" direction.

The essential, profound, "right" fact is that living in the material world with inner harmony and service to others while simultaneously aspiring to make the divine connection are simply flip sides of the same coin. Put another way, human beings are supposed to learn to cultivate inner harmony and at the same time strive to connect with each other to make the earth a better place to live for all. This multi-strata work, undertaken daily, proceeds, however, without too much fanfare simply because it is the "right" thing to do, not because reward is expected.

These very messages, replete in all of the sacred literature, have been glossed over in preference for the empty dogma of religions and material hankerings of human beings that actually mean little in the realms of seriously higher awareness. The work referred to here is really important and requires discipline to accomplish. Even those who understand what ought to be done have a difficult time doing it consistently-- simultaneously disciplining and harmonizing themselves and working to make the world a better place.

A classic little book in metaphysics from the eastern perspective, *The Secret of the Golden Flower*, comments cogently on these profound, essential missions to be accomplished in tandem because one done without the other doesn't amount to much: "Mastery of the inner world, with relative contempt for the outer, must inevitably lead to great catastrophes. Mastery of the outer world, to the exclusion of the inner, delivers us over to the demonic forces of the latter and keeps us barbaric despite all outward forms of culture."[9] Pay attention, and you will uncover numerous examples of barbarism in the world today as in centuries past, the horror recycling generation after generation even among so-called highly civilized and educated people. Simply recall the lack of civility in our daily lives. Warmongering, hate crimes and domestic abuse are on the rise worldwide. And because inner quiet and harmony are also indispensable to higher consciousness, the noise and negative vibrations emitted from over-use of cell phones, computers, iPods and, of course, high tech weapons hamper human beings from connection to the Source. Before he died, controversial writer and social activist Norman Mailer made a frontal assault on organized religion and forged a new catechism that links contemporary technology to weaponry of the devil.[10] Similarly, Russian Daniel Andreev writing secretively while in a gulag in the 1940s also concluded that so much of what new technology offers is an abomination that will be used ultimately to oppress and hurt unsuspecting citizens worldwide.

Big governments and religions have failed to take responsibility for bringing human beings together, and it is becoming increasingly obvious that walls have been engineered to separate us. Propaganda, dogma and zealotry impede the right and good work that must be done to raise collective consciousness, with chaos and tension the rule rather than the exception. At times it seems there is little hope of ever knowing universal harmony. Having been socialized and taught for thousands of years to prefer fighting and loathing to caring about each other, we allow discord, war, hatred, and selfishness perpetuated by governments, religions and a few powerful individuals to pass unprotested. Why? Because bridge-building toward peace and concord brings less profit. Arms dealers are having a grand old time working with officials and their associates to make money and control, oppress and even punish citizens for no high or divine purpose other than greed.[11] Propaganda and dogma are skillfully used by a few high leaders to maintain a status quo that is divisive, anti-democratic, self-serving and more and more counterproductive, creating dissension that goes on and on, far beyond individuals' lifetimes. 'My God is better than your God and I know him better than you do and will proselytize, oppress you or even kill you to prove it' seems to be the way of the world.

How much better off we would be demanding peace, cooperation and love. Ironically, former U.S. President and high ranking World War II officer Dwight D. Eisenhower is often remembered having said, "People want peace so badly that sooner or later governments will have to move over and let them have it." We're still waiting. …

Despite this, some intrepid persons persevere. For instance, Russian aristocrat and Crimean War officer Leo Tolstoy, who also realized the importance of peace and cooperation, wrote about these powerful ideas in *The Kingdom of God Is Within You*. Tolstoy explained that when "war breaks out, in six months the generals have destroyed the work of twenty years of effort, of patience and of genius." Having seen war's horrors, Tolstoy made clear his conviction that it created wretched, horrible situations and, moreover, that it resulted from the "most hideous materialism."[12] Whether Tolstoy embraced the rose symbolically or not, he seemed to know something of its real essence as did his compatriot Daniel Andreev.

In America, Civil Rights leader Martin Luther King, Jr. also well understood the rose's messages, although he is not ostensibly linked to the flower in any particular literature. Dr. King often used mountains as a metaphor in speeches such as "I Have a Dream." The mountain alluded to human beings reaching a higher vantage point from where the full picture is revealed: that we are all connected, all part of the same Source. This, too, is the missive of the rose: it also signals interconnectedness and potential for higher consciousness.

Sadly, Dr. King was an exception. Too many churches and denominations foster not peace but discord. History shows how much blood curdling violence has been committed in the name of religion. As Mark Lilla, professor of humanities at NYC's Columbia University reminds us in a *New York Times Magazine* article, theology is nothing but "a set of reasons people give themselves for the way things are ...."[13] i.e. with too many unscrupulous, avaricious leaders dividing people rather than bringing them together.[14] As English philosopher Thomas Hobbes wrote: "Messianic theology eventually breeds messianic politics,"[15] end-of-days, other-worldly rhetoric designed to link war with salvation rather than with improving THIS world. As Lilla reminds us, religious fervor feeds war, obscuring the real heart of doctrine, that we "must change our lives." This unheeded message, at the core of all major religions and metaphysical movements, seems accessible to few. In other words, few heed the message of the rose.

The rose over centuries of literature

Yet, unusual, mystical qualities of the rose – as inspirational symbol, aesthetic enhancement, curative medicine and nourishing food –, have appeared over centuries and in diverse geographic locations, as Monika Joshi, for instance, shows in Indian and Hindu culture and Michael Price in Native American traditions. Italian poet Dante's acknowledgement of the flower as a sublime component of his initiation in *The Divine Comedy* and Rumi's magnificent use of the rose as a metaphor for cosmic consciousness in his 13th century Persian poetry are other extraordinary examples of the flower's global pervasiveness.[16] Pointing further to mysterious aspects of the rose, James Gaffarel in 1650 noted that it had even an "astral-light" body that actually outlived the life of the flower. It is not clear what prompted Gaffarel to experiment with the flower's aura, but he found something startling: when the ashes of a burnt rose were preserved and held over a lighted candle, the mass became a dark cloud, still in the shape of a rose, "so Faire, so Fresh, and so Perfect a one, that you would have thought it to have been as Substantial and Odoriferous a rose as grows on a rose-tree."[17] While we know nothing of Gaffarel's state of mind or level of higher awareness 500 years ago, his experience with the rose is not unique.

At various times and places, others, too, around the world have also had remarkable rose encounters.

More recently, the purest essential oil of the rose has been recognized as possibly the only natural oil essence that can raise the electromagnetic frequency of human beings to 320 MHz. Human beings rally around 58 MHz while frequencies lower than 32 imply the body's shutting down, i.e. death. Scientist, educator and former Protestant minister Dr. David Stewart may be the first in modern times to teach the power of rose oil. The author of *The Chemistry of Essential Oils Made Simple*, Dr. Stewart notes that, in comparison to others, the oil of the rose contains 11% alkanes, an important hydrocarbon, and other properties that render it unique.[18] Granted that the alkanes of the rose have properties not fully understood by chemists, these contain nonetheless "spiritual and healing qualities ... recognized and applied for thousands of years" (Stewart). Thus pure rose oil when anointed with good intent is thought to be precious and indispensable.

Perhaps even more mysterious than Gaffarel's astral-light experience with the rose or Dr. Stewart's findings about the chemical components of rose petals is a series of paranormal episodes experienced by a physician and dentist, Dr. John Ballou Newbrough during the last years of the nineteenth century. Born in Ohio in 1828, Newbrough attended Cincinnati Medical College and advanced quickly as a result of a superior mind and training, first in medicine and then in dentistry. He was also gifted in the paranormal from teenage years onward. Gaining financial independence during the California 1849 gold rush, Newbrough married the sister of a prospecting friend and settled in New York City where he practiced medicine and connected with a spiritualist group. Newbrough's clairvoyant and clairaudient powers were sharpened as he cultivated "out of body intelligences," but by his own admission he remained "disgusted with the low grade of intelligence displayed by them."[19] In other words, it appears that Newbrough wanted to advance faster than his spiritual guides were leading him, a fact also true of Elizabeth Haich who admits in her book *Initiation* that she begged for help of a master to advance in cosmic consciousness.[20] Apparently some human beings are chosen for initiation to higher realms; others seek initiation but are denied. Why this is the case is not easily understood but it can be deduced through the study of metaphysics that past life karma, akashic records, and sincere heartfelt intent all have a bearing on illumination. So too do hard work, service, silence and discipline of one's body, including diet. Nevertheless, between 1871 and 1881, while living in New York City in the vicinity of what is now Pennsylvania Station, Dr. Newbrough is said to have received "spiritual guidance" toward purification to the extent that he was told to stop eating meat, to discipline himself and to purchase a typewriter, an instrument that had just been introduced to the marketplace, which he did. As the little known Newbrough story goes:

> Upon sitting at the instrument an hour before dawn he discovered that his hands typed without his conscious control. In fact he was not aware of what his hands typed unless he read what was being printed. He was told that he was to write a book - but must not read what he was writing until it was completed. At the end of a year when the manuscript was completed he was instructed to read and publish the book titled *OAHSPE*, a new Bible.[21]

With guided hands, what Dr. Newbrough produced from that new typewriter was a strange, absolutely remarkable bible known by a select few as *OAHSPE*. It consisted of 36 books, over 800 pages, a number of illustrations and pictographs, and was a provocative rendering to say the least. One of the books or chapters in *OAHSPE* known as "Cosmogony and Prophecy" tells "what light, heat, electricity, gravity, etc. [are] and what causes them, what holds planets in their places, gives the many cycles of time used by the ancients in their tables of prophecy" and more. Another segment, "The Book of Jehovih's Kingdom on Earth" is cogent in its instructions to human beings who want to "develop spiritual powers, prophetic abilities and extra-sensory perception." One warning is particularly cogent: to stop eating blood-thirsty carnivorous animals and stick to "herbivorous foods."[22] Prophetically, "The Book of Jehovih's Kingdom on Earth" mentions "a disease [that] came among the cows and the physicians forbade the babes being fed on their milk." Instead, he recommends corn and rice milk, "an excellent liquid food for infants."[23] Otherwise, the same book admonishes humans to know the importance of learning: "And ye know that all light is progressive. Ye cannot settle down, saying I know enough." "The Book of Discipline" instructs humans how to treat each other: *very well*, and discipline in all matters is a key aspect of this urged quantum link. A verse in "The Book of Discipline" is cogent in its message that connection to cosmic consciousness follows abandoning "earthly habits and desires" and "constantly putting away the conditions below."[24]

Now, much in *OAHSPE* is so other-worldly as to be indecipherable but many pages make good sense, and the rose metaphor appears in a number of short passages; some books even contain pictographs of a rose, most notably in "The Book of Saphah" dealing with the Tablet of Hy'yi:

> Behold, the rose, deep rooted in the earth, Jehovih riseth in majesty of All Light. His colors no man maketh, nor knoweth any man the cause. This subtle Perfume, whence cometh it, and whither goeth it? What power fashioneth it and propelleth it?[25]

The message concerning the rose rooted deep in the earth has multiple meanings, two of which inhere in this essay. The first and most important alludes to the human potential to open unseen chakras or energy centers of the body.[26] These chakras unfurl like a flower when people perform continual sincere work on themselves, provide service to others and show overwhelming pure intent to reach Cosmic Consciousness (or what Protestants know as Christ Consciousness). People who have accomplished some level of work have a glow or radiance about them that is sometimes

Tablet Of Hy'yi

noticed by others but tends not to be missed by those who have worked on higher awareness. Unevolved humans in the presence of these special "light" beings simply note that they are "different." What's more, such individuals avoid the fray and foolishness of material life.

The rose, then, with its roots, stem, thorns, and petals models the work and rewards of the spiritual search. Furthermore, as we have seen, its medicinal properties and alluring scent are similarly precious, and these play a major role in *OAHSPE*. The pictorial referred to as the Tablet of Hy'yi in "The Book of Saphah" inspires this conclusion. Perhaps decipherable only by readers immersed in metaphysics or theosophy, the interpretation given here may be the first ever to be found in print.

Turning to page 618, plate 81 of *OAHSPE*, you will find a very large rose at the bottom of a pictograph; at the top of the same illustration is a crown adorned with a cross. A rectangle, the pictograph shows two hearts just at the tip of the rose, and a circle with a single straight line drawn through it is nestled between the two hearts.

The same circle symbol is also seen in Helena Blavatsky's *Secret Doctrine* and is explained there in Volume II, page 30, as connected to the root-race of the "sweat-born." Above the circle symbol are an eye and a diamond that in all probability signify the light or ultimate perfection that a human being comes to know through the practice of pure unconditional love on earth, essential to rise up, incorporeally, to celestial heights. Rising up is indicated in Blavatsky's pictograph by what is above the moon and star: a crown connoting cosmic consciousness. Now, a perfect diamond, as in the pictograph, is referred to by the diamond industry as a paragon; it contains 100 carats or more. Thus, when this perfect state is reached, the rose, too, is fully opened because it has found light. In other words, when a human being works really hard to overcome the material world, to love and to serve others, and to develop a sense of community, that person may rise up to connect with cosmic consciousness.

Continuing this theme in Volume II *(Anthropogensis)* of *The Secret Doctrine*, Blavatsky explains that humans may drop the external shape surrounding them and cease to be of the material world – and hence become absolutely Divine. There is enough evidence that this process can and does take place on earth as in the Holy Spirit mysteries connected with the lives of Jesus Christ, Mithra and others. There are among us masters yet and still, but their presence can be sensed only by those in true service who are deeply committed to the work that must be done. In this regard, a number of groups today including the Eckankar movement and even some governments are interested in the process of rising up or detaching from one's body, also controversially known as astral travel or remote viewing, as those in the military internationally call it.[27] While it has been proven that some evolved human beings actually have this capability, astral travel does not necessarily confer illumination or full initiation into Cosmic Consciousness, the so-called eternal life. Much has been written about the process of rising up to the realm of the masters, but extremely few are invited into this region. By all accounts, to move in this direction requires extraordinary discipline, inner harmony and good intent toward others. Discord and lack of good works are sure ways not to be connected to higher consciousness.

A number of additional verses in *OAHSPE* connect the rose to this striving for perfection. "The Book of Apollo," for instance, beseeches readers to "Behold the rose and the lily; they are perfect in their order. Being one with Jehovih, they painted not themselves." This book advises the "symmetry of flesh; the symmetry of spirit; the harmony of music, [and that humans should] consider wisely their behavior." Once again, this message is contained in practically all sacred literature although it is one of the most unheeded revelations in the universe.

A paradoxical footnote to the work of Dr. Newbrough is his effort to establish a utopian community known as Shalam near Las Cruces, New Mexico, in 1884 in order to teach others to practice many of the precepts of perfection outlined in *OASPHE*. All-encompassing for Newbrough and his friend Andrew Howland, Shalam was one of many utopian communities begun in America in the 19th century. Howland and Newbrough put their life savings into the community dedicated to training a diversity of orphaned children to be disciplined and spiritually exceptional. The successes were remarkable as described by Lee Priestly who has researched the Shalam community:

> Naturally the spiritual development of the children was of prime importance. In addition to the study of *OAHSPE*, they

were taught the means of communicating with angels and how to discriminate between good and evil spirits. At the ages of twelve to fourteen they were initiated into the rites and ceremonies of the Ancients with explanation of appropriate signs, symbols and emblems. Trances and manifestations were common-place to them. They received spirit messages and responded to rappings and table tippings.[28]

As Priestly notes, the children, as focal point of the community, were also taught manners and decorum. Trained to be unfailingly polite but not forward, they conversed with adults easily and with poise. Among themselves they were loving and noncompetitive.[29] Unfortunately, the same cannot be said of the adults of the community known as Faithists. Not surprisingly, the grown-ups at Shalam were described by Priestly as men and women who shirked labor, expected something for nothing, lounged idly and were known generally to be quarrelsome and critical; in other words, their comportment was typical of human beings who have not come to understand the work on themselves that is absolutely necessary for higher awareness.

At age 63 in April 1891, Newbrough and a number of the children at Shalam died of influenza. With Howland at the helm, the Shalam community lumbered along for over a decade after Newbrough's death but was beset by law suits and bickering among the resident adults. Although Howland was said to be fair, reasonable, and progressive in introducing many new vegetables and fruits to the area, he was not the manager of the caliber that Newbrough had been. Noted for his strange attire – white pantaloons and a long flowing white beard –, Howland was arrested twice for "indecent exposure."[30] Shalam folded in 1907.

Perhaps stranger than fiction or mere coincidence, Helena Petrovna Blavatsky, also telepathic and one of the founders of the worldwide Theosophical movement, died the same year as Dr. Newbrough in 1891. As noted earlier, Newbrough was born in Ohio in 1828 while his contemporary, Blavatsky, was born in Russia in 1831. There is no evidence that the two knew each other. They traveled in different circles, he in the United States and Australia and she in India, the United States and all over Central Europe where *The Secret Doctrine* was penned.

The evidence is that *The Secret Doctrine*, well over 2600 pages, was spiritually dictated to Blavatsky while she resided in several locations including in Elberfeld, Germany; Ostend, Belgium; and around England with Countess Constance Wachtmeister as her companion a good portion of the time. The two were frequently in the company of visitors from the Theosophical Society who were witnesses to some of the unusual circumstances surrounding Blavatsky's work on the opus.[31] By Blavatsky's own account she was visited by an ascended master in 1851 while touring Hyde Park in London with her father for the first time; she was then quite young, only 17 years old.[32] It is generally thought that Blavatsky's guide was Master Morya, the head of "all of the esoteric schools which truly prepare an aspirant for ashramic contact and work."[33] But long before 1851 there had been considerable drama and mystery surrounding the privileged Russian woman known as HPB among theosophists. To add to her mystique, she is known to have run away from a seventy-five year old retired military husband at age 17 after declaring herself a widow who "wouldn't be a slave to God Himself, let alone man."[34] Known to be hot-tempered, Blavatsky was, by all accounts, an oddity. Those who came into her orbit were likely to witness some very strange, inexplicable things whenever she was around; by all accounts she was much stranger than her contemporary Dr. Newbrough. Yet, precisely because she was so remarkably unusual, stories about her are kept alive.

A number of people aware of her "gifts" tracked her from an early age including key members of her family. Known as Helene von Hahn at birth, in childhood HPB she was already recognized as a medium and became especially known for very high intellect in the years after her mother's death in 1842. She was also "the strangest girl one has ever seen, one with a distinct dual nature ... one mischievous, combative, and obstinate – everyway graceless; the other as mystical, metaphysically inclined. ... No schoolboy was ever more uncontrollable or full of the most unimaginable pranks ..."[35] Manly Hall's insightful article about her in *The Phoenix* benefited from the journal notes of HPB's aunt in Russia:

> It was as though a troop of sprites were in constant attendance, ever mindful of her bidding. Her uncanny prophetic powers by turns amazed and discomfited visitors. ... Looking

them intently in the face, like some Pythoness or Delphi, in half-formed childish words she would solemnly predict the place and time of their death. … [H]er prognostications were so correct that she became the terror of the domestic circle. Nothing could be concealed from her that she desired to know. She could read the most hidden thoughts and motives and was constantly aware of circumstances occurring at a distance.[36]

Dr. Newbrough, it may be recalled, was also approached psychically in a preliminary visit from a master, but much later in his life than HPB and closer to the time he received through dictation his one and only opus, *OAHSPE*. There is no evidence that either HPB or Newbrough was fearful or apprehensive about undertaking the tasks given them by masters, but other differences exist. Blavatsky, for instance, wasn't asked to alter her diet or to purchase a typewriter. Other than both being "visited" years in advance of the major tomes they were guided to write, the only other noteworthy similarity is that both Newbrough and HPB were plagued by scandal and charged by "outsiders" as impostors. It could be reasoned that Blavatsky's reputation suffered most from scandal since her work came to the attention of theosophists, non-theosophists and even the global press which followed her work rather closely. Attempts to discredit HPB are not surprising since she was guided by an otherworldly source that few human beings could comprehend then or now. As Manly Hall explained in an early issue of *The Phoenix*, "Humanity attacks viciously and relentlessly anyone who assails the infallibility of the mediocre." What's more, "The fear is not that the occultist may be wrong; the fear is that the occultist may be right."

Ironically, the information dictated to Blavatsky contained two major divisions known as "Cosmogenesis" (Volume I) and "Anthropogenesis" (Volume II) which are also aspects of the voluminous revelations dictated to Newbrough.

Notwithstanding that style, terminology and organization of these two authors' books differ greatly, and *OAHSPE* contains far more illustrations than *The Secret Doctrine*, both are major works that the masses have overlooked. It is pure conjecture to suggest that both oeuvres have been reserved for or intended for a select few, but that would appear to be the case. For is it mere coincidence that the symbol of the rose appears quietly in both *OAHSPE* and in *The Secret Doctrine?* Or that the two authors, born only three years apart, died the same year after having been given major Arcanum or keys to understanding the cosmological foundations of the universe? What Newbrough and Blavatsky provided far exceeds the cosmological and anthropological insights presented in any other religious or sacred literature, yet few are aware of their work.

The Rose in the Emerald Tablets

Regarding Blavatsky's importance, let's first turn to the presence of Atlantis in her writing in order to work our way back to the rose. Blavatsky knew that the so-called lost continent had indeed existed, because she understood "axial disturbances" as integral to the "intelligent Kosmic hand and law which alone could reasonably explain such sudden changes... [as the disappearance of Atlantis]." According to HPB's *Secret Doctrine*, "old continents were sucked in by the oceans, other lands appeared and huge mountain chains arose where there had been none before." Plato, who supplies additional evidence for this in *Timaeus*,[37] joins HPB who interprets the very large statues found in the Easter Islands and in the Gobi Desert as gesturing toward a lost continent, the Atlantic Ocean portion of Lemuria where human-like beings over "nine yatis" or "27 feet tall" could be found. In HPB's view, this was the cradle of the third root race that occupied a vast area of the Pacific and Indian oceans.[38] Thirty-five years ago, not coincidentally, Erich von Daniken depicted the same large statues in his controversial film "Chariots of the Gods."[39] If, then, Atlantis existed, let's concede that the giants residing there left evidence of their intelligence in the statues reminiscent of themselves made of a rock substance that has lasted for millennia. Nor should it surprise us if at least some of this highly evolved Atlantean tribe lived by a set of laws set down in a special, little-known book called *The Emerald Tablets*. Their leader, whose home was known as Keor, was named Thoth, an eternal being who knew the secrets of everlasting life and led a group of thirteen adepts who were said to be well versed in science and philosophy.[40]

In recent times, renewed interest in *The Emerald Tablets* has raised conjecture about the origin of this important work, including who fetched it back to the Pyramids in Egypt. Michael Doreal is said to have revived and translated the *Tablets*. Thought by some to reincarnate the great teacher Horlet who lived in Egypt during the time of Atlantis, at least 20,000 years ago, Doreal was a great if not well-known spiritual leader, healer and teacher who was born Claude Dodgin in Oklahoma, USA.[41] Although his life events appear nowhere near as controversial or as stupefying as those connected to Newbrough or HPB, there were some spiritual oddities, most notably the fact that he was "given" the task of retrieving the mysterious *Emerald Tablets* from the jungles of the Yucatan in a Temple of the Sun God and returning them to the Great Pyramid in Egypt. Through great peril, Doreal is said to have translated and returned the sacred, secret *Tablets* to the designated place; under his direction the book was published in English in 1939. Doreal also taught higher awareness in metaphysics courses for a number of years to a select group of males and females in Oklahoma[42] and is best known as the leader of the American wing of the Temple of the White Brotherhood founded in the Himalayas.

Now, *The Emerald Tablets* often mentions a rose-like flower that has petals, most likely a rose – in any case, as in other sacred literature, an impeccable, revered flower. Planted in the human body, invisible buds bloom along a high-level 'yellow brick road' to cosmic consciousness, a path that seekers uncover through service, love and inner harmony. Notwithstanding life's trials and tribulations, the flowers point toward opportunity to find the light when the journey ends, the shared veiled essence of many classic stories, including *The Wizard of Oz*, *The Magic Flute* and much esteemed literature that a broad readership can enjoy even if the ultimate meaning is clear to only a few. *The Emerald Tablets*, although eons old and recondite, still have much to teach about this blossoming: "In freeing consciousness from the body it is best to expand the solar-plexus (one of the chakras), the Flower of Life of the body, and send the life force flooding through it so that the body is vitalized in preparation for consciousness to leave…"[43] With hard work, good intentions, and silence as privileged gardeners, *The Emerald Tablets* underscores an important fact about higher consciousness: "He who talks does not know; he who knows does not talk." For the highest knowledge is "unutterable."

Similarly, the rose makes its quiet appearance in the King James version of the Christian *Holy Bible* as well as in Gnostic portions such as the *Book of Enoch* that have been omitted, for whatever reason, from the King James version. Thought to predate the New Testament, *The Book of Enoch* provides powerful rose metaphors such as the birth of Methuselah's strange son whose "body was white as snow and red as the blooming of a rose..."[44] Undoubtedly this metaphor references Methuselah's illumination or connection to the Holy Spirit. By now it is apparent that the Bible has been altered numerous times[45] and a significant part, the Gnostic portion, has been hidden from the masses. Still, the allegories, parables and metaphors contained in this revered book, if truly understood, are instructive and, to say the least, full of revelations.

Because not based on historical truth or intended for literal consumption, the Bible, when read and quoted, does not easily release its many meanings. Indeed, the beauties and enigmas of the *Holy Bible* derive from its openness to multiple interpretations. The *Holy Bible*, moreover, contains hidden knowledge, in parable, allegory and metaphor not intended for readers unprepared for it; indeed, many passages, for good reason, have been cut. Geoffrey Hodson, well known as a teacher and leader in theosophy, explains carefully in *The Hidden Wisdom in the Holy Bible*, Volume I, that there is much support for a symbolic reading of the Bible in, for instance, "the promises of perpetual prosperity and divine protection made by God to Abraham and his successors with subsequent defeats by invaders, exile under their commands in Babylon and Egypt and destruction of the Temples of King Solomon." More recently, the author of *Rescuing the Bible from Fundamentalism, Born of a Woman* and other significant works, Episcopal Bishop John Selby Spong has made clear his view that even theologians have misunderstood and misinterpreted much of the *Bible*. This proviso notwithstanding, we find two perhaps obscure but absolutely beautiful references to the rose in the Old Testament: in "Song of Songs," chapter 2, and in "Isaiah," chapter 35, verses that are easy to miss because often misinterpreted and disputed.

"Isaiah" is said to be the most prophetic book of the Old Testament, revealing unique prophecies regarding Immanuel and the "Suffering Servant." The prophet Isaiah is also said to have been a literary genius. Of all the prophets he looked further into the future than any other. According to some sources, he lived a long time, through the reign of three kings, and was even an advisor to one of them, Hezekiah, from 729 to 699 B. C.

The 35th Chapter of "Isaiah," verse one, uses the rose metaphor. It reads: "the desert shall rejoice, and blossom as the rose." Interpreted, the literary trope means after the necessary work has been completed, tribulation will cease and blessings will flow. Obvious to astute readers, the verse has nothing to do with a desert per se. It means that "all spiritual evil and physical catastrophes will be reversed and the land and people will be blessed."[46] In fact, the book of Isaiah is replete with this same message in veil: potentially, we may all connect with the Holy Spirit if we reverse course, slay the ego, care for one another, and work fully toward the light. When this is accomplished even the 'tongue of the dumb will sing', to use another metaphor for rising up out of the material realm which brings everlasting joy and no 'sorrow and sighing'.

A bevy of Biblical scholars has interpreted the "Song of Songs" as an allegory, a typology, an anthology of love songs, a three-character interpretation and a literal love story, but the overall purpose is "to show the joy of married love as a gift of a good and loving God."[47] All of the interpretations seem plausible, but my feeling is that none gets at the real hidden message in this most beautifully written Old Testament book. Like countless other ambiguous biblical verses, those that concern the rose have multiple meanings, an intended opacity because, like the sages and mystics before him, Jesus taught in parables. Fully allegorical and metaphoric, the *Bible* offers sacred, arcane messages intended for those few willing to work to mine for them.[48]

Let's consider the work involved in understanding "Song of Songs," interpreted as an erotic verse or even a song about two lovers, the shepherd-king and his beloved outcast maiden. The story goes that after a period of absence, the former shepherd Solomon returns as King and takes the teen away with him in a royal coach to become his bride. This quiet verse is reminiscent of Rumi's poetry. The essential point lies in the way Solomon and his beloved embrace each other, in verse 13 of chapter one, for instance, when she wants him "betwixt her breasts." But there is nothing carnal in her wishes. 'Between her breasts', a metaphor, means transcending this world, shedding the duality so rooted in human nature, or, for students of the "Law of the Three in Sacred Geometry," abandoning the polarity of the material world. In Taoist tradition, getting between a woman's breasts is similarly non-erotic, transcending the yin and the yang or dual nature of humanity to shed the material world. As human beings move toward objective consciousness, higher and higher levels can be achieved, usually over lifetimes reaching up to merge with the Divine. Thus, in "Song of Songs," if the King gets between, that is, beyond the maiden's breasts, he will receive what is better than all the "other daughters" can possibly give him, a chance to transcend the material world. Erotic on the surface, the verse strongly implies release. Chapter 2, verses one, two and three, use the rose as a powerful metaphor; in them, the female steadfastly declares that she is superlative:

> I am the Rose of Sharon and the lily of the valley, so is my love among the daughters as the apple tree among the trees of the wood, so is my beloved among sons. I sat down under his shadow with great delight, and his fruit was sweet to my taste.

If we read this most beautiful verse of the Old Testament in today's vernacular, we would find a well built, fantastic "knock out" who has it "going on" to such an extent that no other "daughter" in the group can measure up to her. This is the case metaphorically when one reaches Cosmic Consciousness. Nothing else measures up. After the connection is made, everything is magical and perfect. Nothing else is desired or needed! One literally and figuratively rises above this material world, above this earth, to connect with the Source. What Dante, Rumi, Hanai, the prophet Isaiah and King Solomon all seem to know, few of us are able to grasp or to do the disciplined work to realize that absolute perfection can be attained if and when we ratchet up to the Holy Spirit. There is no denying that the Spirit is the key but it takes work to connect with it.[48] Sadly, every day, millions upon millions attend mosques and churches, perform masses and other religious rituals, yet have not come to understand this fact. The kind of attendance needed is attendance to oneself – to be a better human being and to connect with respect and kindness to each other.

More about the unique rose

Much has already been revealed about the rose, yet other dimensions of this quintessential flower can be named. Perhaps no symbolism is more striking and possibly confusing than its connection to Parseeism, the worship of deified fire. This is not fire in a literal sense, according to Hargrave Jennings, but rather "the inexpressible something of which real fire or rather its flower" is symbolized in the rose.

Fire worship is an odd belief system and philosophy that rarely makes its way even into philosophy textbooks yet has been a component of pagan and primitive worship for thousands of years and continues in some traditions even today. Strange as it may seem, this connection to fire is one of the early justifications for the thousand-year-old practice of cremation.[49] Even the torches used at early funerals were not solely for light. Torches, light, candles and other sources of illumination when used in sacred services are rather the "inexpressible mystery of the Holy Ghost."

Fire, like roses, has had a strange history in religion and metaphysics. Combustion's connection to religion, metaphysics and God-seeking in general has actually been stranger than fiction. The world over, for thousands of years, the yearning to connect to God or achieve cosmic consciousness has bred macabre efforts rooted in misunderstanding the meaning of fire in connection to religion and God-seeking. People burned animals, themselves and each other as sacrificial offerings to curry favor with God or their gods.[50] At times self-immolation or sacrifice has been practiced in Indian and Tibetan Buddhism and in some sects of the Middle East, the whole purpose of which was to leave the material body. All of these sacrificial rituals have been unnecessary and a perversion of what was intended. While fire symbolically relates to Spirit, undeniably, it is not to be taken literally; self-immolation profits one nothing. The point of the best theology, too often veiled, is to learn to live together in harmony and also to find inner peace. When these two goals can be accomplished in tandem, a light appears and gets brighter and brighter – that is, an invisible internal flower opens, and the worshipper potentially becomes one with Spirit.

The *Holy Bible* (St. James version) is replete with fire metaphors and allegories, most notably in "Daniel": "a fiery stream issued and came forth before him…" Thus, God manifested in fire or Spirit. Now, the book of Matthew warns that baptism is generally performed with water, but eventually "He shall baptize you with the power and glory and with fire." What kind of God would appear as fire, promise to baptize with fire and not use water? Fire philosophers had a ready answer: a God who is one with the Holy Spirit and is the Source of All and Everything. As Michael Doreal clarifies in his interpretation of *The Emerald Tablets*, "Man's destiny is the final blending with light even though he moves through darkness during material incarnations."[51] This explains why some cultures prefer cremation and why burial ceremonies rely on fire. It is also, sadly for the misguided, one reason for self-immolation. Thus the "signification of fire burial is the commitment of human mortality into the last of all matter, overleaping the intermediate states; or delivering over of the man-unit into the Flame –Soul, past all intervening spheres or stages of the purgatorial."[52] Fire symbolizes the Holy Spirit and, as we have seen, the rose symbolizes this as well; among the blossom's many meanings, none is as urgent and important as its connection to fire as symbolic of the Divine. When the magnificent rose opens to the light, or when petals of all of the invisible chakras in the human body open to the light of the harmonious and good, we find the symbol of connection with the Holy Spirit or Cosmic Consciousness.

In his writings and teaching, the great philosopher, alchemist, and physician Philip Theosophratus Aureolous Bombast Hohenheim, widely and controversially known as Paracelsus (1493-1541) understood the connection of fire to the Holy Spirit. To say the least, Paracelsus did not live a charmed life. But his was so exciting and unusual that stories have been written about him, including one by Argentine novelist Jorge Luis Borges (1889-1986) titled "The Rose of Paracelsus." In the Borges' story, when a doctor (Paracelsus) asks God for a disciple, one appears. But the fellow is full of doubt; he asks for evidence of the great doctor's power. Paracelsus listens intently to the disciple's request for proof, namely that the master turn a rose into ashes and then make it reappear. Paracelsus burns the rose but doesn't renew it. Alas, the would-be disciple leaves; not having witnessed revival, he remains unconvinced that he should follow Paracelsus. After the disciple exits, however, Paracelsus quietly creates a fresh rose from the ashes. The moral of the story seems to be that if humans doubt the power of the rose, the power of the Holy Spirit, then its benefits are not for them and the connection cannot be made.

Yet, the rose has a democratic spirit; it can become the province of all. An efficient astrologist, alchemist and philosopher, Paracelsus was also a preacher who, addressing crowds in taverns, mixed religious and political polemic that called for social equality based on Christian principles.[53] In words that are as apropos today as during the Renaissance years in which he lived, Paracelsus warned, "No good can happen to the poor with the rich being what they are. They are bound together as with a chain. Learn you rich, to respect these chains. If you break your link, you will be cast aside."[54] We cannot deduce from Paracelsus' teaching that he favored the poor over the rich, but he advocated cooperation between social classes and ethnicities, a melding of aims for the benefit of all. Ultimately, such collaboration follows the spirit of the rose. It is wise for us to mimic this spirit and be ever mindful of the Source.

References

[1] Hans Biedermann, *Dictionary of Symbolism: Cultural Icons and the Meanings behind Them.* (NY: Penguin, 1992), 290.

[2] Marcell Jankovics, *Book of the Sun.* Trans. and ed. Mario Fenyo (Wayne, NJ: Hungarian Studies Publications, Inc. and Columbia UP, 2001), 110.

[3] Donald Tyson, ed. *Three Books of Occult Philosophy: Written by Henry Cornelius, Completely Annotated with Modern Commentary; The Foundation Book of Western Occultism.* (St. Paul, MN: np, 2004), 132.

[4] For more on the rose nebula, see Jerry Bonnell and Robert J. Nemiroff, *Astronomy 365 Days: The Best of the Astronomy Picture of the Day website.* (NY: Harry Abrams, 2006). Also of interest is *Constellations, Stars and Celestial Objects* (Stuttgart, Germany: Firefly Books, 2005).

[5] Rüdiger Dahlke, *Mandalas of the World: A Meditating and Painting Guide* (NY: Sterling Publishing, 2001) 136.

[6] In the Christian *Holy Bible's* numerous translations and versions, not all elect to use the rose in metaphor, but the rose appears in the books of "Isaiah" and "Song of Songs" in the most widely read King James version. In the book of "Isaiah," Chapter 35, *The Holy Bible: New International Version* makes reference to the crocus instead of the rose. Interestingly, the King James version is thought by some to have been written by Francis Bacon on behalf of King James I of England; this notion is feasible since Bacon was steeped in Masonic and metaphysical traditions and is even thought to be the true author of Shakespeare's plays and sonnets that are also replete with parable and allegory that reference the divine.

[7] Israel Regardie, *The Golden Dawn: A Complete Course in Practical Ceremonial Magic* (St. Paul, MN: Llewellyn Publications, 1986), 47.

[8] Joseph Needleman, *Why Can't We Be Good?* (NY: Penguin, 2007), 252.

[9] Richard Wilhelm, *The Secret of the Golden Flower: A Chinese Book of Life.* Trans. Richard Wilhelm with Commentary by Dr. C. J. Jung. (NY: Harcourt Brace Jovanovich, 1931), viii.

[10] See Norman Mailer, *On God.* (NY: Random House, 2007).

[11] To corroborate this notion of war, weapons and conflict at the bidding of govern-

ment, see, for instance, *Blackwater USA*, 2007 that explains the development of a private military force.

[12] Leo Tolstoy, *The Kingdom of God is Within You: Christianity Not as a Mystic Religion But as a New Theory of Life*. Ed. Martin Green. (Lincoln, NE: University of Nebraska Press, 1984), 152.

[13] Mark Lilla, "The Politics of God," *New York Times Magazine*, August 19, 2007, 30.

[14] In this regard, See Jim Marr, *The Secret War*. (NY: Random House, 2003) and Anna Politkovskaya. *Putin's Russia* (London: The Harville Press, 2004).

[15] Mark Lilla, "The Politics of God," 50.

[16] Of interest is a symphony produced in Rumi's honor by Hafez Nazeri during this 800th year celebration of his birth and the United Nations declaration of 2007 as the International Year of Rumi. In September 2007, the University of Maryland's Center for Persian Studies, College Park, MD sponsored a 3-day event honoring the work of Rumi. See *The Washington Post*, August 30, 2007, C6.

[17] As quoted in Manly Hall, *The Teachings of All Ages*. (New York, Penguin Group, 2003), 347.

[18] David Stewart, *The Chemistry of Essential Oils: God's Love Manifest in Molecules*. (Marble Hill, MO., Care Publications, 2006), 153.

[19] John Ballou Newbrough, *OAHSPE, A New Bible in the Words of Jehovih and His Angel Ambassadors*. (London: Kosmon Press, 1942), iv.

[20] See Elisabeth Haich's account in *Initiation*. (Santa Fe, NM: Aurora Press, 2000).

[21] *OAHSPE*, v.

[22] *OAHSPE*, Book of Jehovih's Kingdom on Earth, Chapter VIII, Verse 8, 814.

[23] *OAHSPE*, Book of Jehovih's Kingdom ..., Chapter XI, Verse 29, 818.

[24] *OAHSPE*, Book of Discipline, Chapter III, Verse 14, 837.

[25] *OAHSPE*, Book of Saphah, 618.

[26] See C.W. Leadbeater. *The Chakras*. (Wheaton, IL: Quest Books, 1927). It is now in its 9th printing. On page 7, Leadbeater notes that there are seven chakras. Also of interest is Ruth White. *Using Your Chakras: A New Approach to Healing Your Life*. (NY: Barnes and Noble, 1998).

[27] Regarding remote viewing, of interest are Paul H. Smith, *Reading the Enemy's Mind: Inside Star Gate*. (NY: Tom Doherty Associates, 2005) and Jim Schnabel, *Remote Viewing: The Secret History of America's Psychic Spies*. (NY: Bantam, Doubleday, Dell Publishing, 1997).

[28] Lee Priestly, *SHALAM: Utopia on the Rio Grande, 1881-1907*. (El Paso, TX: Western Press, 1988), 32.

[29] Ibid., 33.

[30] Ibid., 22.

[31] Countess Constance Wachtmeister et al., *Reminiscences of H. Blavatsky and the Secret Doctrine*. (Wheaton, IL: Theosophical Publishing House, 1976).

[32] Master DK or Djwhal Khul is said to have been HPB's guide or master. Of interest regarding ascended masters is Phillip Lindsay. *Masters of the Seven Rays: Their Past Lives and Reappearance*. (Queensland, Australia: Apollo Publishing, 2006), 49.

[33] Ibid., 41.

[34] Mary K. Neff, ed. *Personal Memoirs of H. Blavatsky*. (NY: E. Dutton & Co., Inc.

1937), 37.

[35] Ibid., 26.

[36] Manly Hall, ed. *The Phoenix: An Illustrated Review of Occultism and Philosophy*. (Los Angeles: Hall Publishing, 1931-32), 86.

[37] Noteworthy is Lewis Spence. *The History of Atlantis*. (London and NY: Rider & Son, 1926). Plato's work, *Timaeus* was published 400 years before the birth of Christ.

[38] See H. Blavatsky, *The Secret Doctrine: The Synthesis of Science, Religion and Philosophy*, Volume II, "Anthropogenesis" (London, Theosophical Publishing Company, 1888), 333.

[39] See Erich von Daniken's film *Chariots of the Gods* released in 1972 by Sun Classic Pictures and packaged in 2005 by Blair and Associates, Ltd. Von Daniken's controversial theory is that man did not descend from apes as maintained by Charles Darwin but that humans have descended from gods.

[40] Michael Doreal, *An Interpretation of the Emerald Tablets Together With The Two Extra Tablets*. (Castle Rock, CO: Brotherhood of the White Temple, n.d.), 8.

[41] Whitby, Delores, *Doreal: As I Knew Him*. (Sedalia, CO: Little Temple Library, 1980), 5.

[42] Ibid., 15.

[43] *The Emerald Tablets*, 132.

[44] R.A. Charles, ed. *The Book of Enoch*. (San Diego, CA: The Book Tree, 1917; rpt. 2006), 91. Another reference appears on page 64.

[45] As many scholars concur, the *Bible* has been altered and translated a number of times. In *The Teachings of All Ages*, page 543, for example, Manly Hall explains that regarding the King James version, the translator Sir Francis Bacon was given the task of checking, editing and revising the Scriptures. According to Hall, the very first edition of the King James version contains a cryptic Baconian headpiece. Interestingly, Phillip Lindsay in *Masters of the Seven Rays* notes that Francis Bacon "wrote secret ciphers ... for the Holy Bible [King James version] and [for] Shakespearean plays." See *Masters*, 29.

[46] *Holy Bible, King James Study Bible*. (Nashville: Thomas Nelson Publishers, 1988), 1010. Interpretations from this version are divided thus: (1) allegory: treats the view of early Jewish literature, denies the literal aspects (2) Typology: sees marriage as a type of Christ and the Church (3) Collection of love songs: understands the Song of Solomon as many love songs with no unified meaning (4) Love Triangle: emphasizes the love triangle or "eternal triangle" with Solomon as the villain who tries to lure the maiden from her shepherd-boyfriend and (5) Literal love story: views this book as just that, a love saga.

[46] Ibid.

[47] Hidden knowledge and the secret teachings of Jesus are mentioned and alluded to throughout the *Bible*, including in Mark 4:11 which says "Unto you it is given to know the Kingdom of God: but unto them that are without, all these things are done in parables." That is, many worldwide read the *Bible*, but few fully understand it. A tremendously complex book, it demands years of study, reflection, and sincere intent to grasp, as we find reiterated in the Book of Mark, Chapter 4, verse 12: "Seeing they may see and not perceive; and hearing they may hear and not understand; lest at any time they should be converted and their sins should be forgiven them."

[48] Ibid.

[49] Beverly Moon, ed. *The Encyclopedia of Archetypical Symbolism: The Archive for Research in Archetypical Symbolism.* (London: Shambhala, 1991), 308.

[50] See Marcell Jankovics, *Book of the Sun.* Trans. Mario Fenyo. (NY: Columbia UP, 2001), 120. Also of interest is Mel Gibson's movie, *Apocalypto* (2006) that graphically depicts the practice of human sacrifice during the 15th century Aztec empire.

[51] Michael Doreal, *An Interpretation of the Emerald Tablets,* 49.

[52] Hargrove Jennings, 111.

[53] Philip Ball, *The Devil's Doctor: Paracelsus and the World of Renaissance Magic and Science.* (NY: Farrar, Straus and Giroux, 2006), 124.

[54] Ibid., 124.

Acknowledgments

As editor, compiler and contributor to this anthology, I find my gratitude is as far reaching and varied as the chapters presented here. Some of my appreciation takes on veiled dimensions that should remain so. However, I thank teachers and colleagues who inspired or aided the project, perhaps in most cases unknowingly. These include Sy Ginsburg, Koschek Swaminathan, Don Conte, Albert Amao, Lillian Mayer, Mary Brandis, Mike Machiopa, Mem Masnick and others. I am also grateful to Leon Fair, Lew Ross, and Leonard Jackson for their "rose insights" and to Vernon O'Meally for overall administrative assistance to advance my chapters. One of my former graduate students, Lisa Cucciniello, contributed more than her own chapter on the rosary in the Catholic tradition; she also provided editorial and research assistance to a couple of other contributors. Likewise, Alexsandr Prodovikov was a faithful research assistant whose service was invaluable and steadfast over the span of the project. The Jacques Marchais Museum of Tibetan Art in Staten Island, New York, made available rare metaphysical books from the private collection of Jacques Marchais that were exceedingly helpful.

On a trip to Osaka in the summer of 2007, I had the good fortune to meet Hisae Ogawa who introduced me to the work of Dr. Tomin Harada while we were touring Japanese cultural sights. The anthology is enriched, it is certain, by my overview of Dr. Harada's medical work to help the victims of the 1945 bombing of Hiroshima and sufferers of agent orange exposure during the Viet Nam War. We applaud the unique rose garden movement begun by Dr. Harada in Japan.

Perhaps it is not inappropriate to also thank my late Dad, Frank Pauling, for he taught me much about the workings of the world in parable and metaphor as well as by his example of hard work, service and belief in family. During my pre-teen and teenage years he reminded me regularly that what I didn't know would "make whole new worlds." Of course, until now, it was not at all clear what he meant by that. The lessons he taught become clearer and clearer with each passing year and every time I see a rose in full bloom. My parents' work to raise their children in the midst of rank racism, to serve their community as best they could, and to stay married amid all sorts of negative outside forces, I can now see was a feat bordering on the phenomenal.

Finally, and most of all, abundant thanks to all who felt the mission of the rose and worked with me over the months, now years, that it took for the project to reach fruition. To all those who contributed essays or assisted directly in the anthology's production, including Albert Amao; Lisa Cucciniello and Alex Prodovikov in New Jersey; Mario Fenyo and Olga Mavlyutoza in Maryland; Monika Joshi in California; Tobe Levin in Germany; Godfrey Okorodus in Belgium; Leon Fair in Nebraska; Michael Price in Minnesota; Montgomery Taylor in New York City; our new contributors Ryan Dunne, Vicki Tuongvi Eaton, Yaping Qian ("Cathy"), Uma M. Swaminathan and Maria Jaschok as well as Dr. Ding Zhanggang ("Dylan") in Beijing, China, for managing the team that translated the Chinese version of *Rose Lore*, I say thank you in the spirit of the rose.

Frankie Pauling Hutton, Ph.D.
2007/ 2012 /2015

Contributors to the 2015 Edition of *Rose Lore*

Albert Amao
Having attended San Marcos University in Lima, Peru, and currently residing in New Jersey, Albert teaches and researches metaphysical topics and authored several books in this field. He lectures nationally on theosophy.

Lisa Cucciniello
A public school teacher and former graduate student, Lisa is an avid researcher and resides in New Jersey, USA.

Gamze Demirel
A classically trained Turkish scholar, Gamze is a former collegiate professor of history at Suleyman University, Istanbul, Turkey.

Ryan Dunne
A Baltimore, Maryland, Irish Tenor, actor and musician, Ryan is a graduate of the Maryland College of Art and the founder and lead musician of "Never Bird." He researches Irish folk music and culture.

Vicki Tuongvi Eaton
A former radio talk show host, Vicki is a native of Vietnam, a sound therapist and wellness coach. She resides in New Jersey, USA.

Maria Jaschok
With a PhD in Chinese Social History, Maria Jaschok directs International Gender Studies at Lady Margaret Hall, University of Oxford and co-founded the Women and Gender in Chinese Studies Network (WAGNet).

Monika Joshi, R.N.
A registered nurse and former officer in the Indian military, Monika now lives in California and is an Ayurvedic practitioner.

Sy Ginsburg
Sy is a Chicago-based attorney and teacher of metaphysics, including facilitating Advaita and Theosophy groups.

Tobe Levin von Gleichen
Founder of UnCUT/VOICES Press devoted to fighting female genital mutilation, Tobe is currently a Visiting Research Fellow in International Gender Studies, Lady Margaret Hall, University of Oxford and an Associate of the Hutchins Center for African and African American research at Harvard University.

Frankie Pauling Hutton
A former journalist and collegiate professor of history, Frankie is the author of several award winning scholarly books, including *The Early Black Press in America* and *Outsiders in 19th Century Press History: Multicultural Perspectives.* She is the founder of the Rose Project, <www.roseproject.com>.

Michael Wassegijig Price
A member of the Wikwemikong First Nations in Ontario, Canada, Michael has been an academic dean at a Tribal College in Minnesota and a consultant to NASA and *National Geographic Magazine.* He resides in Montana, USA.

Yaping Qian/ "Cathy"
A collegiate professor at China Women's University, Beijing, China, Cathy teaches in the English department and is assistant director of the Translation Center at CWU. Her original research on the rose in Chinese history and culture was done exclusively for *Rose Lore.*

Uma M. Swaminathan
A collegiately trained anthropologist and businesswoman, Uma is also a filmmaker, artist and Reiki Master. She grew up in the Himalayas and divides her time between South India and New Jersey.

Montgomery "Monty" Taylor
Monty is a New York City talk show host and master astrologer who regularly teaches and conducts workshops in New York.

www.ingramcontent.com/pod-product-compliance
Lightning Source LLC
Chambersburg PA
CBHW030405270326
41926CB00009B/1272